Resistance to Innovation

Resistance to Innovation

Its Sources and Manifestations

SHAUL OREG & JACOB GOLDENBERG

THE UNIVERSITY OF CHICAGO PRESS   CHICAGO AND LONDON

SHAUL OREG is associate professor of organizational behavior at the School of Business Administration at the Hebrew University of Jerusalem. He is a coeditor of *The Psychology of Organizational Change*. JACOB GOLDENBERG is professor of marketing at the Arison School of Business at the Interdisciplinary Center Herzliya, visiting professor at Columbia Business School, and the author or coauthor of several books, including *Inside the Box*.

The University of Chicago Press, Chicago 60637
The University of Chicago Press, Ltd., London

Printed in the United States of America

24 23 22 21 20 19 18 17 16 15   1 2 3 4 5

ISBN-13: 978-0-226-63260-5 (cloth)
ISBN-13: 978-0-226-23732-9 (e-book)
DOI: 10.7208/chicago/9780226237329.001.0001

Library of Congress Cataloging-in-Publication Data

Oreg, Shaul, 1970– author.
Resistance to innovation : its sources and manifestations / Shaul Oreg, Jacob Goldenberg.
pages cm
Includes bibliographical references and index.
ISBN 978-0-226-63260-5 (hardcover : alk. paper)—ISBN 978-0-226-23732-9 (e-book)
1. Technological innovations—Social aspects. 2. Technological innovations—Psychological aspects. I. Goldenberg, Jacob, 1962– author. II. Title.
HD45.O7294 2015
658.5'75—dc23

2014033480

♾ This paper meets the requirements of ANSI/NISO Z39.48–1992 (Permanence of Paper).

Contents

Introduction

New product introduction is a major activity of firms. About 25,000 new products are introduced each year in the United States alone. Most of them, however, end up failing (McMath and Forbes 1998; Bobrow and Shafer 1987). In view of the fact that expenditures for developing new products increase as the marketing process advances toward the product's launch, it is critical for firms to screen out concepts and ideas that are likely to fail, as early in the process as possible (Dolan 1993). Previous research on new product performance has shown that a wide variety of factors influences the outcome of new product development activities (cf. Montoya-Weiss and Calantone 1994; Freeman 1982; Virany, Tushman, and Romanelli 1992; Cooper and Kleinschmidt 1987; Lilien and Yoon 1989). These determinants usually involve some combination of strategy; development process; and organizational, environmental, and market factors. But despite the large body of research on success and failure, product failure is still surprisingly common. Whether one defines failure as the product not reaching expected sales, or as an absolute low level of sales, the large majority of products end up failing, with 95% failing according to the former definition, and 60% according to the latter.

Aiming to maximize performance, marketers in the field of innovation focus their attention on consumers who react favorably to innovations and have little resistance, if any, toward them. We don't blame them; these are the people who are first to adopt innovations, they read consumer reports, participate in online innovation-oriented forums, and even provide companies with feedback on their products. In an innovation's life cycle, they are the first to appear, and their adoption is imperative for products' success. Everett Rogers (2003) identified two such groups of consumers: "Innovators" and "Early Adopters." The other consumers constitute the large majority, who, marketers believe, will eventually adopt because progress is inevitable, and Innovators will "help" them see the value in the innovation. Although recently more attention is given to the majority, we cannot ignore the fact that the voice of Innovators is much more readily available on Internet forums, exhibitions, stores, and even market surveys than that of the majority. This corresponds with an overall pro-change bias by both researchers and practitioners (Ram 1987; Schwarz and Shulman 2007; Sheth 1981).

If we consider early adopters as high on innovativeness, what does that make of the rest? According to the current view, this majority, which includes approximately 80% of the consumer population (Rogers 2003), is currently characterized by what it *lacks* (innovativeness), rather than what it *possesses*. Such a characterization hinders a true understanding of these consumers' attitudes and behaviors and of the diverse reasons they may have for resisting an innovation. From a practical perspective, lacking this understanding, marketers fail to identify potential means that will make adoption easier for these 80%. In fact, a common assumption is that consumers with particularly high levels of resistance should be of little interest to firms given the low revenue they generate. In this book we suggest the contrary, and that rather than focusing on what is absent from most consumers, we would be wise to focus on what is present: Varying degrees of *Resistance to Innovation*.

Furthermore, using a definition that focuses on what consumers lack is unjustified academically in the same way we do not (or at least should not) aim to prove a null hypothesis. The marketing literature has not established that the majority lacks innovativeness, nor, as the philosophical discussion refers to it, can one prove that an orange crow does not exist. Empirically, one can only prove that which exists. Part I of this book is dedicated to understanding resistance and its sources.

Contrary to popular thought, resistance does not inevitably lead to product failure, just as innovativeness does not inevitably lead to product success. Product failures are only one of several ways in which resistance is manifested. In part II we show that resistance is not an inherently negative concept—just as not all new innovations are inherently beneficial for individuals. In fact, resistance is often manifested for products that ultimately succeed.

One of our aims in this book is to bring together multiple views on resistance to innovation and demonstrate their joint implications. The examples we discuss demonstrate that resistance to innovation is more than the opposite of "innovativeness." It is the outcome of an interplay of personality traits and other, context-specific, factors and mechanisms. Given the combination of factors that influence people's response to innovation, this response is often complex. People may often be ambivalent about an innovation, both captivated and threatened by it. As we discuss in detail in chapter 2, for example, this is the case for many technological advances that require individuals to relinquish their hard-earned skills in using earlier technologies. Others may simply be misinformed or deterred by the manner in which the innovation is presented.

In contrast to the scant attention the term has received in marketing, the term *resistance* is widely used in the organizational and management literatures to describe employees' responses to organizational changes. Based on the conceptual foundations established by Kurt Lewin in social psychology (1935, 1947), Lester Coch and John French conducted in 1948 what was perhaps the first empirical study of resistance to change. Since then, scores of studies in organizational behavior have been devoted to this subject, with an ongoing momentum of studies today. These studies typically seek to explain resistance by exploring its sources. Research in marketing, however, has more typically considered the market *consequences*, rather than the *sources* of consumers' resistance.

Note, however, that whereas the study of resistance to change in organizations is more restrictive than the study of resistance to innovation by the mere fact that the former is limited to organizations and the latter is not, resistance to innovation is a more restrictive phenomenon given that the introduction of an innovation is but one type of change. In this book we integrate insights from these two bodies of literature to provide a more complete depiction of the resistance phenomenon, involving both its sources and consequences.

Throughout the book, we use the term *innovation* in reference to "an idea, practice, or object perceived as new" (Rogers 2003, xx), and define the innovation decision process as "the process through which an individual (or other decision-making unit) passes from gaining initial knowledge of an innovation, to forming an attitude toward the innovation, to making a decision to adopt or reject, to implementation of the new idea, and finally to confirmation of this decision" (Rogers 2003, 168). The reasons for, and consequences of, resistance to innovation that we discuss are not tied to a particular form of innovation and rather revolve around the notions of novelty and uncertainty that characterize all innovations. Therefore, although most of the innovations we discuss in the book have to do with technological advances, the insights we provide about resistance to innovation can be readily translated to other forms of innovation, such as other types of products, ideas, or procedures.

Defining Resistance

Merging insights from organizational behavior and marketing requires that we first address how the term *resistance* is defined in each. In the organizational literature, resistance to change is typically defined as a negative attitude, comprising affective, cognitive, and intentional components (Oreg 2006; Oreg, Vakola, and Armenakis 2011; Piderit 2000). Individuals who resist change may experience anxiety when learning about the change, they may rationally evaluate the change as being bad, and they may explicitly object to the change, speak against it among colleagues (that is, generate negative word of mouth), and make active efforts to prevent it. The same definition has also been used for changes involving the introduction of new technologies in organizations, in which case the anxiety, negative evaluation, and/or active resistance are directed against the new technology, or mode of work. In the organizational behavior literature, resistance is thus defined on the basis of the ultimate reactions to the change, *independent of the factors that bring about this reaction*.

In contrast, the term *resistance to innovation* in marketing implicitly alludes to the antecedents of the reaction to the innovation. Resistance in this literature implies that irrational factors are responsible for individuals' refraining from adopting a worthy innovation. Stated otherwise, refraining from adopting an innovation because the innovation offers little benefit would not be considered an act of resistance by most marketers. In the organizational field, however, negative attitudes toward organizational changes are viewed as resistance even when the reasons for these negative attitudes may seem trivial, as in the case of a change that is perceived as harmful. We believe that this latter, more inclusive, definition has a number of benefits.

First, it clearly distinguishes the ultimate act of rejecting an innovation from the reasons one may have for doing so. This is important because before one can turn to explaining a phenomenon, it should first be clear what it is that one wishes to explain. Second, determining which reasons are rational and which are irrational may become quite messy. Defining resistance through individuals' reaction toward the innovation leaves this mess outside the definition of the construct and thus provides a more coherent depiction of what resistance is. Related to this point, an unworthy innovation for one individual may be a most valuable innovation for another. Certainly, even utter failures were seen as having utility by those who introduced them and it is only in retrospect that one can more objectively claim that an innovation was truly unworthy, after observing a consensus among consumers who have chosen not to purchase it. And even then, there are cases in which a rejected innovation resurfaces years later, and is then embraced wholeheartedly. Early after launch, however, when understanding resistance may be of most importance, it is too soon to evaluate the perceived utility versus threat of the innovation. It is therefore of value to restrict the definition of resistance to the act of refraining from adoption, and to separately investigate why it is that people resist. Finally, as we unravel in the book, we suggest that even alleged irrational reasons for resisting may ultimately lead to resistance through their impact on the more "rational" perceived benefit versus threat. We propose that such perceived benefit versus threat serves as the most proximal predictor of resistance, mediating the effects of more distal, and often inconspicuous, predictors, such as the innovation's introduction process, or the context in which the innovation is introduced.

The Focus and Structure of the Book

We approach the resistance phenomenon from two main fronts. In the first part of the book we aim to understand its sources. What causes resistance? What are the conditions in which it is most likely to evolve? Or in other words, why and when do individuals resist innovation? Note that we said "individuals." Although what is ultimately of interest to marketers is market resistance, in trying to understand where such market resistance comes from one's focus necessarily shifts to the individual, because, as we demonstrate in part II of the book, market resistance, in its various forms, begins with the resistance of individuals (see figure 0.1). We therefore begin by considering the factors that lead individuals to resist innovations.

In each of the chapters in part I we focus on a different source of resistance. Much of what we suggest is based on what has been learned about resistance to organizational change. This is because of the much larger pool of studies that have been conducted in the organizational field on resistance to change than that of marketing-based studies on resistance to

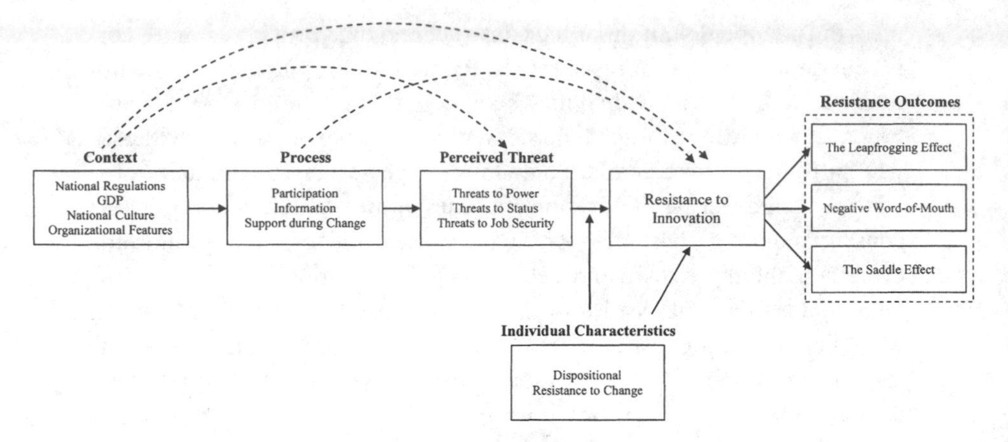

FIGURE 0.1. Antecedents and consequences of resistance to innovation.

innovation. As we will discuss, many of the reasons employees have for resisting organizational change can just as well explain why consumers resist innovations. Moreover, many of the changes resisted in organizations involve the introduction of innovations, and are thus directly revealing for our inquiry of resistance to innovation.

The sources of resistance we consider are drawn from Oreg's and his colleagues' earlier work on resistance to change (Oreg 2006; Oreg 2003; Oreg, Bartunek, and Lee 2014; Oreg, Vakola, and Armenakis 2011; van Dam, Oreg, and Schyns 2008). Specifically, in Oreg (2006) and Oreg, Vakola, and Armenakis (2011), the literature on resistance to change was reviewed and sources of resistance were classified into categories. Oreg, Vakola, and Armenakis (2011) reviewed sixty years of empirical studies of reactions to change and through an inductive process proposed five resistance antecedent categories, of which we believe four are highly relevant for explaining resistance to innovation. These are *individual characteristics*, *perceived threat*, *the innovation's introduction process*, and *the innovation's context*. More recently, the interrelationships among these antecedents have been conceptually denoted (Oreg, Bartunek, and Lee 2014). Each of the first four chapters of the book is devoted to one of these categories, and in each we describe cases that depict resistance due to the corresponding antecedent category. We review relevant literature in the field of organizational change and draw insights framed for our specific interest in innovations.

By classifying variables that antecede resistance into broad antecedent categories, our model (see figure 0.1) offers a comprehensive yet still relatively parsimonious description of the resistance process. Furthermore, beyond integrating previous findings about resistance antecedents and applying them to the innovation adoption context, we proceed in our model and propose how resistance antecedents may relate to one another vis-à-vis their relationship with individuals' resistance to innovation. Specifically, in several of the chapters we provide new empirical evidence for the interrelationships among resistance antecedents that have been previously established only theoretically (Oreg, Bartunek, and Lee 2014).

In chapter 1 we discuss the personality factors that have been shown to contribute to consumers' resistance to innovation. A key factor that explains resistance to innovation lies within the consumer. Consumers differ from one another in their overall orientation toward the adoption of new products. Whereas some are keen to be the first to adopt (a.k.a. dispositional "Innovators"), others are predisposed to resist innovations (a.k.a. dispositional "Laggards").

Beyond individuals' personal inclinations, innovations are often resisted because of what they entail. Adopting the innovation may incur costs, often making resistance not only likely but also rational. As we discuss in chapter 2, this is often the case for innovations in the organizational context, where changes may threaten individuals' status, expertise, or even job security.

In chapter 3 we consider the manner, or process, in which the innovation is introduced, as an antecedent of resistance. Factors within the innovation process include the amount of information that is provided about an innovation and the extent to which prospective consumers, both within organizations and in the marketplace as a whole, are involved in the development and introduction stages. As we suggest in the chapter, these factors can go a long way in explaining why even innovations that could clearly benefit users are often resisted.

Finally, in chapter 4 we go beyond the content or process of the innovation and focus on the particular context in which innovations are presented. The environment in which an innovation is introduced may also be a crucial factor in influencing adoption, increasing or decreasing resistance. The degree to which prospective adopters have faith in the innovation agent, the inclination to adopt or resist among the people in one's close environment, and the overall social and cultural environment in which one encounters the innovation are all possible sources of resistance.

In part II we turn to discuss what resistance looks like at the aggregate level and discuss the consequences of resistance for firms and consumers. Resistance can develop in unexpected ways and can spread about in a silent but most consequential manner. As we elaborate in each of the chapters in part II, to evaluate the effect of resistance one needs to adopt a macro perspective and conduct aggregate analyses. The reason is that even if we understand individual-level resistance, the overall market resistance is not necessarily the sum of individuals' resistances. What makes resistance disseminate is the process through which individuals interact. Our focus in this part of the book is therefore on a variety of effects that resistors have on markets.

Specifically, we examine cases in which innovations did not fail altogether, but have not fulfilled their potential because of, resistance, at least at moderate levels. Cases such as these allow managers and marketers space for maneuvering and turning a potential failure to a success. While complete failures should be analyzed as a means of improving the introduction of future innovations, partial failures leave room for real-time interventions that could improve the ultimate diffusion of the innovation. As we will discuss, even when a thorough analysis is conducted for identifying sources of resistance in advance, some level of resistance is still likely to remain. Managers and marketers should therefore be attentive to the inception of resistance and be prepared to intervene early on if product diffusion is to be maximized. As can be seen on the right side of figure 0.1, our discussion in part II of the book will be of three particular outcomes, or manifestations, of resistance to innovation: the *leapfrogging effect*, *negative word of mouth*, and the *saddle effect*.

In chapter 5 we revisit the lagging phenomenon, which constitutes an extreme form of resistance to innovation and describe how, in some situations, and for some products, individuals who are typically predisposed to being Laggards, may skip product generations and appear as Innovators or Early Adopters on product adoption curves. We then argue that these resistors can be valuable to the firm and generate a considerable income. Contrary to common practice, we therefore argue that firms should devote the time to targeting them. We demonstrate how their adoption can substantially increase the NPV (Net Present Value) of the innovation.

In chapter 6 we explore the market-level nature of resistance and the damage for which resistance is responsible. Namely, we discuss the various mechanisms through which negative word of mouth is disseminated and quantify the cost of resistance for firms.

In chapter 7, we present the Saddle phenomenon. Most innovation markets (either consumer or organizational) are dual markets, including a small (consisting of approximately 16% of the market) group of early adopters and a majority of individuals who exhibit some form of resistance. As we'll discuss in the chapter, this resistance results in a dual market structure, which often leads to a severe drop in sales (more than 20%), lasting for two to seven years, before recovery.

Taken together, we believe that the conceptual frameworks we present in the book, along with the empirical evidence we offer for them, provide well-needed structure to extant knowledge on the resistance phenomenon in general and resistance to innovation in particular. Although we discuss relevant implications of our arguments throughout the book, we do not provide explicit recipes for how to deal with resistance. Our interest is on highlighting and analyzing the various mechanisms that explain resistance and its manifestations. Furthermore, the complexity and multifaceted nature of resistance precludes any simple recommendation that could be generalizable across cases. We do, however, link our arguments to specific examples and cases as a first step in bridging theory and practice.

References

Bobrow, E. E., and D. W. Shafer. 1987. *Pioneering New Products: A Market Survival Guide*. Homewood, IL: Dow Jones-Irwin.

Coch, Lester, and John R. P. French Jr. 1948. "Overcoming Resistance to Change." *Human Relations* 1: 512–32.

Cooper, G. Robert, and Elko J. Kleinschmidt. 1987 "New Products: What Separates Winners from Losers," *Journal of Product Innovation Management* 4 (3): 169–84.

Dolan, J. Robert. 1993. *Managing the New Product Development Process*. Reading, MA: Addison-Wesley.

Lewin, Kurt. 1935. *A Dynamic Theory of Personality*. New York: McGraw-Hill.

———. 1947. "Frontiers in Group Dynamics: Concept, Method, and Reality in Social Science; Social Equilibria and Social Change." *Human Relations* 1: 5–41.

Freeman, Christopher, 1982. *The Economics of Industrial Innovation*. Cambridge, MA: MIT Press.

Lilien, Gary L., and Eunsang Yoon. 1989. "Determinants of New Industrial Product Performance: A Strategic Re-Examination of the Empirical Literature." *IEEE Trans. Engineering Management* 36 (February): 3–10.

McMath, Robert M., and Thom Forbes. 1998. *What Were They Thinking?* New York: Random House.

Montoya-Weiss, Mitzi, and Roger Calantone. 1994. "Determinants of New Product Performance: A Review and meta-Analysis." *Journal of Product Innovation Management* 11 (5): 397–417.

Oreg, Shaul. 2003. "Resistance to Change: Developing an Individual Differences Measure." *Journal of Applied Psychology* 88: 680–93.

———. 2006. "Personality, Context, and Resistance to Organizational Change." *European Journal of Work and Organizational Psychology* 15: 73–101.

Oreg, Shaul, Jean M. Bartunek, and Gayoung Lee. 2014. "A Process Model of Change Recipients' Change Proactivity." Paper presented at the 74th meeting of the Academy of Management, Philadelphia, August 2014.

Oreg, Shaul, Maria Vakola, and Achilles A. Armenakis. 2011. "Change Recipients' Reactions to Organizational Change: A Sixty-Year Review of Quantitative Studies." *Journal of Applied Behavioral Science* 47 (4): 461–524.

Piderit, Sandy Kirstin. 2000. "Rethinking Resistance and Recognizing Ambivalence: A Multidimensional View of Attitudes toward an Organizational Change." *Academy of Management Review* 25 (4): 783–94.

Ram, S. 1987. "A Model of Innovation Resistance." *Advances in Consumer Research* 14 (1): 208–12.

Rogers, Everett M. 2003. *Diffusion of Innovations*. 5th ed. New York: Free Press.

Schwarz, G. M., and A. D. Shulman. 2007. "The Patterning of Limited Structural Change." *Journal of Organizational Change Management* 20 (6): 829–46.

Sheth, J. N. 1981. "Psychology of Innovation Resistance: The Less Developed Concept." *Research in Marketing* 4: 273–82.

van Dam, Karen, Shaul Oreg, and Birgit Schyns. 2008. "Daily Work Contexts and Resistance to Organizational Change: The Role of Leader-Member Exchange, Perceived Development Climate, and Change Process Quality." *Applied Psychology: An International Review* 57 (2): 313–34.

Virany, Beverly, Michael L. Tushman, and Elaine Romanelli. 1992. "Executive Succession and Organization Outcomes in Turbulent Environments." *Organization Science* 3 (1): 72–92.

PART I
Sources of Resistance

CHAPTER ONE

It's Not the Innovation, It's the Adopter

Why Some People Are More Likely Than Others to Resist

Individuals' unreasonable commitment to archaic practices is not uncommon. Let us describe a personal experience from the perspective of the first author (S. O.): Until a few years ago, a dear colleague would still do his work using a 1999 computer with an Intel 80486 microprocessor (a.k.a. the i486, which was the predecessor of the Pentium I chip). He used a fourteen-inch tube monitor, the storage devices on his computer were restricted to floppy disks, and obviously the computer lacked USB ports. As a faculty member in the department, he could have easily asked the IT staff to replace his computer with a newer one, yet he chose not to. He liked his computer. He was used to it, and he believed it comfortably served his needs.

I couldn't understand how he could get any work done on it. He had to undertake complicated maneuvers to transfer content from his computer to those of others, he couldn't install newer versions of software, and overall, the computer was extremely slow. I kept trying to convince him that he'd be better off with a new system, but he was not to be convinced. For him, as long as he managed to get his old software to work, he didn't seem to mind about the rest, even if this required him to acquire peculiar rituals and to restart his computer every so often. It seemed clear that there was something more going on. Rather than lacking the need for a new computer, it was clear that other factors predisposed him to stick so adamantly to his old one. When telling Jacob (my coauthor) this story, Jacob realized that he too has a close colleague, a leading marketing scholar, who exhibits similar behaviors. One of his prime research interests is innovation adoption, yet he is proud to be the last person at his university who has used transparences and an overhead projector, until he was practically forced to switch to PowerPoint presentations. Furthermore, much like the case above, his computer was sufficiently old to be the only one to survive a virus attack that infected all of the computers at his school. He is proud of both stories and exemplifies the strategic (and proud) Laggard.

It is therefore apparent that one reason for resistance lies within the individual. As we discuss in the following chapters, alongside its advantages, the adoption of an innovation typically has drawbacks, with numerous reasons to resist it. At the same time, responses to any given innovation vary widely across individuals. Some people are more inclined than others to resist the mere notion of change. Such a predisposition to resist becomes most apparent at the extremes, once the majority of individuals has already made the transition to the new situation, and stands bewildered when faced with those who insist on holding on to the past.

It was cases such as those presented above that piqued my interest in the internal factors that drive some individuals to resist the mere notion of change. At that time, the literature on individuals' resistance to change and innovation was not abundant. Most works discussed resistance from a macro perspective, focusing on the behavior of organizations or markets rather than that of individuals (e.g., Hannan and Freeman 1984). There were some works, however, that looked at the adoption of innovation and change as a function of individuals' personality. Most of these works focused on identifying those who are early to adopt (Innovators) versus those who resist innovations (Laggards). For example, Everett Rogers (1995) proposed that early adopters are empathetic, rational, and well able to cope with uncertainty. David Midgley and Grahame Dowling (1978; 1993) added to these attributes and suggested that Innovators are also relatively high on traits such as achievement seeking and self-monitoring.

Empirical research has been conducted from this perspective and has linked early adoption to traits such as novelty seeking (Manning, Bearden, and Madden 1995); tolerance for ambiguity, low cognitive rigidity, and low dogmatism (Jacoby 1971; Raju 1980); and cognitive innovativeness (Goldsmith, Freiden, and Eastman 1995; Im, Bayus, and Mason 2003; Marcati, Guido, and Peluso 2008; see table 1.1). This research was motivated by the desire to understand the psychological mechanisms that account for early adoption. While we can be relatively certain that Laggards, typically defined as those within the market who are last to adopt

TABLE 1.1 **Personality constructs that have been used for predicting early adoption**

Construct	Works contributing to the establishment of the construct	Definition
Novelty seeking	(Hirschman 1980; Pearson 1970)	The tendency to approach versus the tendency to avoid novel experiences.
Tolerance for ambiguity	(Budner 1962; Frenkel-Brunswik 1948)	The degree to which an individual perceives ambiguous stimuli as desirable, challenging, and interesting, without ignoring or avoiding their complexity.
Cognitive rigidity	(Pally 1955; Werner 1946)	Stiffness, of difficulty in responding efficiently and adequately to changing stimuli.
Dogmatism	(Rokeach 1960)	The degree to which a person's belief system is controlled by the need to defend against anxiety or threat versus the need to know and understand.
Cognitive innovativeness	(Venkatraman 1990)	The preference for engaging in new experiences with the objective of stimulating the mind.

an innovation (Rogers 1995), differ in their characteristics from Innovators, who are first to adopt, we still lack a framework that focuses on Lagging. In line with this view, more than twenty years ago Midgley and Dowling (1993) noted: "It is interesting that as a field we believe the rejection of new products to be common, yet there are few studies that directly address this phenomenon. Clearly research on rejection is needed and is likely to have a high payoff in terms of improving our models" (623–24). Although some exceptions exist, as in a study linking consumers' self-efficacy to resistance to technological innovations (Ellen, Bearden, and Sharma 1991), Midgley and Dowling's criticism is as relevant today as it was then. It is true for research on Laggards in general and for the dispositional perspective of Laggards in particular. Rather than say what Laggards are not, it would be valuable to say what they *are*. Trying to explain why some individuals resist change and innovation may yield insights that otherwise escape us when we focus solely on explaining early adoption and the eager pursuit of change. Now this is not to say we cannot infer characteristics of the dispositional Laggard from studies of Innovators, yet as we will describe shortly, several of Laggards' characteristics are *notably* more than the mere opposite of the characteristics of Innovators.

I (S. O.) therefore began my investigation of the resistance to change concept, with an ultimate goal of empirically establishing a construct describing individuals' dispositional inclination to resist change. What are the factors that lead some individuals to resist the changes and innovations that others readily embrace? How can we conceptualize the "resistant individual"? What is this individual like? To empirically explore these issues, this investigation followed a rigorous set of scale-development procedures to establish the dispositional resistance to change construct and its corresponding measurement scale. These procedures are described in detail below.

Development of the Resistance to Change (RTC) Scale

Item Generation and Scale Administration

To start off in the investigation, I prepared an exploratory study, with the aim of establishing the construct's dimensionality. I then developed the measurement scale following prescriptions from DeVellis (1991) and Hinkin (1998) for item generation, survey design, item reduction, and determination of internal consistency. Based on the available literature on individuals' reactions to change, I generated items to tap each of six conceptual categories:

1. *Reluctance to lose control.* A number of researchers emphasized the role of control and its loss as the primary cause of resistance (e.g., Conner 1992), stating that in many situations individuals will resist the imposition of change, which involves the removal of control over their life situation. Correspondingly, organizational studies that advocate employee involvement and participation in organizational decision making as a means of overcoming resistance to change (e.g., Coch and French 1948; Sagie and Koslowsky 2000) seem to be focusing on this source of resistance.
2. *Cognitive rigidity*. Other researchers have undertaken a cognitive perspective on the resistance to change phenomenon, arguing that resistance is a result of individuals' internal schema regarding the notion of change (Bartunek, Lacey, and Wood. 1992; Bartunek and Moch 1987; Lau and Woodman 1995). The concept of dogmatism (Rokeach 1960) nicely captures such a source of resistance. Dogmatic individuals are characterized by rigidity and closed-mindedness and are therefore less willing and able to adjust to new situations.
3. *Lack of psychological resilience.* A third source of resistance has to do with individuals' resilience and evaluation of the self. Resilient individuals may be less reluctant to initiate changes in their lives because to do so is to admit that past practices may have been faulty. Such willingness to potentially lose face

(e.g., Kanter 1985; Zaltman and Duncan 1977) requires that the individual feel confident enough about him- or herself to admit possible mistakes. Furthermore, considering the added stress that change situations often incur, personal resilience is necessary for maintaining adequate levels of performance during change. Corresponding with such a rationale, resilient individuals were found more willing to participate in an organizational change (Wanberg and Banas 2000) and exhibited improved coping with it (Judge et al. 1999).

4. *Intolerance to the adjustment period involved in change.* A distinct aspect of individuals' psychological resilience is their ability to adjust to new situations. Some have argued that people resist change because of the added work it entails in the short term (Kanter 1985). As we will discuss later (see chapter 2 on the role of perceived threats), new tasks typically require learning and adjustment, and some individuals are more willing and able to endure this adjustment period than others. Those with a greater reluctance to endure this period are more likely to exhibit resistance to change.
5. *Preference for low levels of stimulation and novelty.* Several studies in psychology addressed concepts such as novelty seeking (Pearson 1970), sensation seeking (Zuckerman and Link 1968), and arousal seeking (Mehrabian and Russell 1973), all of which pertain to individual differences in the degree of stimulation people desire and seek. In a highly cited study, Raju (1980) demonstrated that individuals' optimum stimulation level is associated with exploratory behavior in the consumer context, such as adoption of innovations, brand switching, and variety seeking. Accordingly, it is reasonable to expect that individuals with lower levels of optimal stimulation would be more likely to resist changes.
6. *Reluctance to give up old habits.* In the context of resistance to organizational change, several theorists propose the reluctance to give up old habits as a common source of resistance to change (e.g., Tichy 1983; Watson 1971). Some explained this reluctance by suggesting that "familiarity breeds comfort" (Harrison 1968; Harrison and Zajonc 1970). This is because new stimuli may render familiar responses incompatible with the new situation. Such incompatibility may produce stress, which can then become associated with the new stimuli. This association is one of the mechanisms through which habits are formed and maintained, which serves as yet another source of resistance to change.

These six categories, each of which constitutes a broad reason for resisting change, were used as the basis for formulating items for the resistance to change scale (Oreg 2003). Items were designed to distinguish those who are inclined to resist change from those who are not. As with the development of many scales, the idea was to provide examples of responses that are typical for the construct being assessed. In the present

TABLE 1.2 **Sample items from initial resistance to change scale item pool and the classification to affective, cognitive, and behavioral dimensions**

Dimension	Item
Affective	When things don't go according to plans, it stresses me out.
	It feels good to have a consistent schedule.
	I would enjoy changing jobs every few years.
	If I were to be informed that there's going to be a significant change regarding the way things are done at work, I would probably feel stressed.
Cognitive	My views are very consistent over time.
	Once I've come to a conclusion, I don't often change my mind.
	I generally consider changes to be a negative thing.
	I don't change my mind easily.
Behavioral	Whenever my life forms a stable routine, I look for ways to change it.[1]
	I like to do the same old things rather than try new and different ones.
	Sometimes I change things in my life simply for the sake of change.[1]
	I like to experience novelty and change in my daily routine.[1]

[1]Negatively worded items.

case, these involved resistant and non-resistant reactions to a variety of change situations. Respondents were asked to rate the degree to which they could identify with each reaction. Because reactions to change can take on many forms, including affective manifestations (for example, feeling anxious about a change), cognitive evaluations (for example, believing that a change is of little value), and behavioral responses (for example, actively resisting the change), it was important to include items tapping each of these reaction forms (see table 1.2).

For each of the six resistance sources mentioned above, four to ten items were created. Four additional items were created to tap individuals' general orientation toward change (for example, "generally, change is good," "I generally dislike changes"). This process resulted in an initial pool of forty-eight items. Five researchers with experience in the scale-development process then reviewed these items. They were asked to assess the appropriateness of the items, looking out for ambiguous wording, double-barreled items, and redundant items. As a result, six items were discarded, two were rephrased, and two new items were generated, reducing the pool to forty-four.

Response options for the items ranged from 1 ("strongly disagree") to 6 ("strongly agree"). Using a snowball sampling technique, which is convenient for reaching a diverse set of respondents, the scale was administered to 226 individuals from a broad range of occupations and personal backgrounds. Fifty-four percent of respondents were men, and respondents' age ranged from eighteen to sixty-seven (M = 31, SD = 13.5).

RTC Scale Dimensionality

Following the removal of eleven items that yielded particularly low correlations (r < 0.4) with all other items (Hinkin 1998), an Exploratory Factor Analysis was conducted, using a Principle Components Analysis with an oblique rotation. The oblique rotation was selected because trait dimensions are conceptually related to one another.[1] Items that did not load significantly on any factor or that similarly loaded on more than one factor were discarded. Ultimately, a four-factor solution was obtained (see table 1.3).

The first factor contained items that pertained to one's preference for routines (for example, "I'll take a routine day over a day full of unexpected

TABLE 1.3 **RTC factor loadings for the final item pool EFA***

Item	Factor Loadings			
	F1	F2	F3	F4
Routine-seeking Eigenvalue of 8.9, 38.7% variance explained				
I'd rather be bored than surprised.	.829[a]			
Generally, change is good[b c]	.826			
I'll take a routine day over a day full of unexpected events any time.	.761			
Whenever my life forms a stable routine, I look for ways to change it[c]	.686			
I prefer having a stable routine to experiencing changes in my life[b]	.569			
I generally consider changes to be a negative thing.	.503			
I like to do the same old things rather than try new and different ones.	.496			
I like to experience novelty and change in my daily routine[b c]	.490			
Emotional-reaction—Eigenvalue of 1.9, 8% variance explained				
If I were to be informed that there's going to be a significant change regarding the way things are done at work, I would probably feel stressed.		.902		
If I were to be informed that there is going to be a change in one of my assignments at work, prior to knowing what the change actually is, it would probably stress me out.[b]		.862		
When I am informed of a change of plans, I tense up a bit.		.699		
When things don't go according to plans, it stresses me out.		.675		

continues

TABLE 1.3 ***(continued)***

Item	Factor Loadings F1	F2	F3	F4
If my boss changed the criteria for evaluating employees, it would probably make me feel uncomfortable even if I thought I'd do just as well without having to do any extra work.		.639		
If in the middle of the work year, I were to be informed that there's going to be a change in the schedule of deadlines, prior to knowing what the change actually is, I would probably presume that the change is for the worse[b]		.633		
Short-term focus—Eigenvalue of 1.3, 5.6% variance explained				
Changing plans seems like a real hassle to me.			.749	
When someone pressures me to change something, I tend to resist it even if I think the change may ultimately benefit me.			.680	
Once I've made plans, I'm not likely to change them.			.444	
Often, I feel a bit uncomfortable even about changes that may potentially improve my life.			.418	
Cognitive-rigidity—Eigenvalue of 1.2, 5% variance explained				
I don't change my mind easily.				.740
I often change my mind[c]				.711
My views are very consistent over time.				.668

*Adapted from Oreg (2003).
[a]Loadings lower than 0.3 are not listed.
[b]These items were ultimately eliminated at the reliability-analysis phase in order to remove redundancy. They did not appear to add substantial theoretical content and were highly correlated with the remaining subscale items.
[c]These items were reverse coded prior to running the analysis.

events any time"). More specifically it involved individuals' behavioral tendency to form and maintain consistent schedules and stable routines in their lives. The items it included derived from the literature on optimal levels of stimulation (Goldsmith 1984; Raju 1980) and habits (Harrison 1968; Watson 1971).

The second factor contained items pertaining to individuals' emotional reactions to change (for example, "When I am informed of a change of plans, I tense up a bit"). It pertained to how individuals typically feel when encountering change, with a specific focus on the degree of stress they experience during change. This factor comprised items pertaining to psychological resilience (e.g., Ashford 1988; Judge et al. 1999) and a reluctance to lose control (Conner 1992).

The third factor consisted of items that tap, in one way or another, the degree to which individuals focus on a change's short-term hassles, versus long-term benefits (for example, "I often feel uncomfortable even about changes that may potentially improve my life"). For each of the items, endorsement suggests a preference for the immediate relief from anxiety over the potential long-term benefits of the change. In a sense, these items involve an irrational form of resistance to change in that they all describe resistance that arises in spite of one's awareness to the change's overall potential benefits. Items in this factor derived from the literature on psychological resilience (e.g., Ashford 1988; Judge et al. 1999) and a reluctance to lose control (Conner 1992).

The fourth factor included three items that pertain to individuals' cognitive flexibility versus rigidity. The three items addressed the ease and frequency with which individuals change their minds (for example, "I don't change my mind easily"). It therefore taps content similar to that tapped by scales of dogmatism (Rokeach 1960) and the need for cognitive closure (Kruglanski 1989).

Each of the scale's factors highlights a different aspect of resistance or a different reason for individuals to resist change. While the RTC dimensions are distinct, they are nevertheless related to one another and are all part of the same overarching construct. The four-dimensional structure suggests that those who typically hold a negative orientation toward change, who hold on to the old ways of doing things, and who tend to avoid initiating change in their lives, do so because they enjoy their routines and feel more comfortable having them; they feel uncomfortable and even threatened when change is imposed upon them; they tend to focus on the short-term inconveniences in change rather than its potential long-term benefits; and they find it difficult to let go of preconceptions and preestablished ideas and plans. However, before these inferences about the trait's dimensionality could be made more conclusively, the findings of the above study needed to be replicated and the scale that emerged required a set of further validation procedures.

Validating the RTC Scale's Structure

A separate set of studies was designed to validate the scale's structure. In two samples, one of university employees and another of undergraduates, Confirmatory Factor Analyses (CFA) supported the four-dimensional structure of the scale (Oreg 2003, studies 2 and 3). More recently, the scale's structure was tested in samples of undergraduates from seventeen countries (Oreg et al. 2008; see table 1.4). This test answers the need to

TABLE 1.4 **Descriptive statistics on samples' demographics from Oreg et al.'s (2008) validation study***

Country	Town	*N*	Language	Religion (majority)	% Female	Mean Age (SD in parentheses)
Australia	Burwood and St. Lucia	251	English	30% atheist	67	21.09 (3.61)
China	Beijing	194	Chinese	—[1]	56	20.72 (1.09)
Croatia	Zagreb	246	Croatian	81% Roman Catholic	83	21.43 (1.79)
Czech Rep.	Brno	224	Czech	50% Roman Catholic	78	22.49 (2.10)
Germany	Braunschweig	206	German	51% Protestant	49	23.03 (4.35)
Greece	Athens	386	Greek	87% Greek Orthodox	60	20.97 (2.31)
Israel	Haifa	241	Hebrew	83% Jewish	82	24.35 (3.21)
Japan	Tsukuba	337	Japanese	—	23	19.71 (1.62)
Lithuania	Vilnius	212	Lithuanian	96% Catholic	77	20.31 (1.67)
Mexico	Mexico City	265	Spanish	82% Catholic	51	20.62 (2.19)
Netherlands	Tilburg	205	Dutch	—	80	20.22 (3.45)
Norway	Bergen	266	Norwegian	67% Christian	74	23.24 (4.40)
Slovakia	Bratislava	171	Slovakian	50% Catholic	54	21.40 (1.10)
Spain	Salamanca	288	Spanish	—	59	21.90 (1.55)
Turkey	Istanbul	241	Turkish	98% Muslim	39	21.04 (1.52)
UK	Durham	204	English	95% Christian	45	19.22 (1.83)
US	Auburn, AL	264	English	49% Christian	50	21.19 (2.38)
Total		4201			60.41	21.35 (2.37)

*Adapted from Oreg et al. (2008)

[1]In some countries it was deemed inappropriate to collect data on respondents' religion in the context of this study. This information is therefore missing for these countries.

validate psychological constructs across cultures. Given that people from different cultures often differ in how they typically think, feel, and behave, before a construct can be validly interpreted in different cultures, it is necessary to ensure that it is perceived in the same way across these cultures.

A first step in establishing the cross-cultural validity of the RTC scale was established by conducting a separate Confirmatory Factor Analysis in each of the samples. As expected, the scale's four-factor structure was established in all seventeen samples (see table 1.5). As can be seen in table 1.5, in each of the samples the CFA fit index reaches satisfactory levels, with the exception of Slovakia where the Goodness of Fit Index was slightly below 0.90.

TABLE 1.5 **Descriptive statistics, scale reliabilities, and confirmatory factor analysis results in the 17 samples in Oreg et al. (2008)***

		Reliabilities (α)										
Country	N	Openness	Conservation	Self-transcendence	Self-enhancement	RTC	RTC Mean	RTC SD	$\chi^2_{(107)}$	RMSEA[1]	CFI[2]	GFI[3]
Australia	251	.78	.73	.80	.85	.82	3.09	.57	172.56	.050	.93	.93
China	194	.82	.75	.81	.85	.85	3.14	.62	170.07	.055	.94	.91
Croatia	246	.82	.84	.83	.84	.84	3.01	.61	159.88	.045	.97	.93
Czech Rep.	224	.85	.83	.87	.86	.84	3.13	.56	184.24	.057	.92	.91
Germany	206	.79	.65	.75	.88	.77	3.12	.48	131.36	.033	.97	.93
Greece	386	.71	.58	.75	.87	.72	3.03	.50	227.29	.054	.93	.94
Israel	241	.80	.81	.80	.84	.85	3.15	.59	193.42	.058	.93	.92
Japan	337	.81	.72	.80	.78	.75	3.22	.52	199.46	.051	.91	.93
Lithuania	212	.76	.81	.82	.87	.77	2.86	.51	171.39	.053	.92	.91
Mexico	265	.71	.76	.80	.78	.79	2.79	.58	216.74	.062	.92	.90
Netherlands	205	.78	.74	.83	.83	.85	3.17	.52	177.59	.058	.94	.91
Norway	266	.79	.76	.73	.86	.84	2.91	.56	218.21	.063	.92	.91
Slovakia	171	.77	.75	.79	.84	.79	3.27	.51	184.28	.065	.90	.89
Spain	288	.79	.76	.84	.82	.81	3.01	.58	165.97	.044	.95	.94
Turkey	241	.74	.77	.80	.83	.77	3.03	.54	188.86	.056	.90	.91
UK	204	.77	.77	.83	.84	.78	3.02	.51	190.22	.062	.90	.90
US	264	.64	.73	.71	.72	.83	3.05	.54	160.90	.044	.95	.94
Mean	247.12	.77	.75	.80	.83	.80	3.06	.55	183.08	.050	.93	.92

*Adapted from Oreg et al. (2008).
[1]Root-mean-square error of approximation.
[2]Comparative fit index.
[3]Goodness of fit index.

Undertaking a more rigorous approach to validating the scale across these samples, we also conducted a Multigroup Confirmatory Factor Analysis. This analysis provided support for the scale's four dimensions, although in three of the countries (Greece, Slovakia, and the United Kingdom), the cognitive-rigidity factor did not load significantly on the higher-order RTC factor. A similar finding was recently provided in samples of employees from Russia and the Ukraine (Stewart et al. 2009, but see also Oreg 2009). Indeed, of the four subscales, cognitive rigidity has consistently yielded weaker correlations with the remaining three subscales. Yet why it should be more explicitly separate from the overarching RTC construct in these particular countries is not clear. We could find no common denominator among the countries in which cognitive rigidity did not load on the higher-order factor and therefore suggest some more in-depth, and perhaps qualitative, examinations of how scale items are perceived in the various countries.

Overall, the distinctiveness of the cognitive rigidity dimension has been explained by suggesting that contrary to the three other RTC dimensions, cognitive rigidity involves strong personal convictions and a form of stubbornness that are typically associated with higher levels of self-confidence (Oreg et al. 2008). Indeed, whereas the other three subscales have been shown to correlate positively with neuroticism and negatively with self-esteem and self-efficacy, cognitive rigidity has shown the reverse pattern of relationships. Therefore, it may be that alongside its change-related content, the cognitive rigidity subscale also taps content that is dissimilar and even contrary to that captured by the remaining subscales. For some reason, this latter content takes precedence over the former in some of the countries we sampled. Nevertheless, given the patterns of relationships found between cognitive rigidity and several external variables (see details in the following section), which correspond with the relationship between the overarching RTC construct and these external variables, we suggest that the subscale "taps a unique yet meaningful portion of variance in individuals' reactions to change" (Oreg et al. 2008, 943) and should therefore be maintained.

Construct Validity—Dispositional Resistance to Change and Other Personality Constructs

To this point, the scale's four-dimensional construct has been established. However, what does the construct's nomological net look like? How does it relate to previously established personality constructs? Does the RTC scale indeed measure what it was designed to measure? To answer these questions, studies were conducted in which RTC was measured together with other personality constructs (Oreg 2003, study 3; Oreg et al. 2008; Saksvik and Hetland 2009).

A good personality system within which it would be interesting to position the dispositional resistance to change trait is the five-factor model of personality (also known as "the Big-Five"; Digman 1990). The model evolved from a series of independent studies, using varying methods, aimed at providing an overarching personality structure. The model brings order to the disorganized array of narrow and specific personality traits that exist in the personality literature by delineating a set of broad personality dimensions that encompass a large portion of the variance in individuals' personality. One of the common means through which the model was established was using the lexical approach (e.g., Goldberg 1982), by which personality-related adjectives are culled from the dictionary and used to create a self-report questionnaire. Once questionnaires are administered, data are factor analyzed, and the resulting solution provides a model for the structure of personality for the given language.

Although there is some dispute regarding the precise number of dimensions to cover the content of personality (e.g., Goldberg 1994; Lee and Ashton 2004), by now numerous studies have been conducted, in numerous languages, most of which have provided support for a five-factor solution (e.g., Goldberg 1990; McCrae and Costa 1987). In most formulations, the five broad personality dimensions depicted are *openness to experience*, *conscientiousness*, *extraversion*, *agreeableness*, and *neuroticism*. Of these, openness to experience, which involves attributes such as creativity, originality, and open-mindedness, would appear to be most closely related to dispositional resistance to change. Those who are open to experience are expected to exhibit relatively low levels of dispositional resistance. Indeed, in two independent studies, a negative correlation was found between RTC and openness to experience (Oreg 2003, study 3; Saksvik and Hetland 2009). In addition, a positive correlation was found between RTC and neuroticism, which reflects a lack of emotional stability and a higher propensity to experience anxiety and depression. Such a finding corresponds with the affective nature of dispositional resistance, and in particular to the emotional reaction dimension. When measured along with narrower and more specific traits (also in Oreg 2003, study 3), RTC was positively correlated with risk aversion (Slovic 1972) and dogmatism (Rokeach 1960) and negatively correlated with sensation seeking (Zuckerman 1994) and tolerance for ambiguity (Budner 1962).

Beyond traits, another personality system—that of personal values—has also been used to validate the RTC scale. Contrary to traits, which describe typical behavior, values describe the outcomes to which people aspire and serve as guiding principles in people's lives (Schwartz 1992). Revealing the correlations between dispositional resistance to change and values will shed further light on the motivations of those who resist change. Among the most rigorously validated value systems is Shalom Schwartz's theory of personal values, which divides values into four broad dimensions. Two of these are of particular relevance to the notion of change and innovation, and include the values of conservation and openness to change. Whereas individuals who value conservation tend to emphasize issues relating to security, conformity, and tradition, those who value openness to change emphasize notions of autonomy and self-direction. As one might expect, dispositional resistance to change has been shown to correlate positively with conservation values and negatively with openness values (Oreg et al. 2008).

Thus comparisons of the RTC scale with previously established personality constructs yield expected relationships and serve to validate the measure. The RTC scale appears to capture a variety of change-related features that, while peripheral in other dispositions, are defining characteristics of dispositional resistance to change. The next step in the scale's validation process was to test whether it could be used to predict individuals' actual reactions to changes and innovations.

Concurrent and Predictive Validity—Predicting Reactions to Actual Changes and Innovations

If the RTC scale indeed captures the essence of dispositional resistance to change, then one should be able to use it to predict how people react when encountering specific change situations. Does it predict individuals' change-related choices? Does the adoption behavior of those who score high on RTC differ from that of those who score low? In the context of innovation adoption, does it, as our model in figure 1.1 suggests, predict individuals' resistance to innovation?

In the context of university life, one of the common change situations students regularly come across is the add/drop period when students may change the schedules of their preselected courses. Classes are typically selected before the beginning of the academic year. During the first week or two of classes, students often do some "shopping around," they check out a variety of classes, and ultimately they may select either to hang on

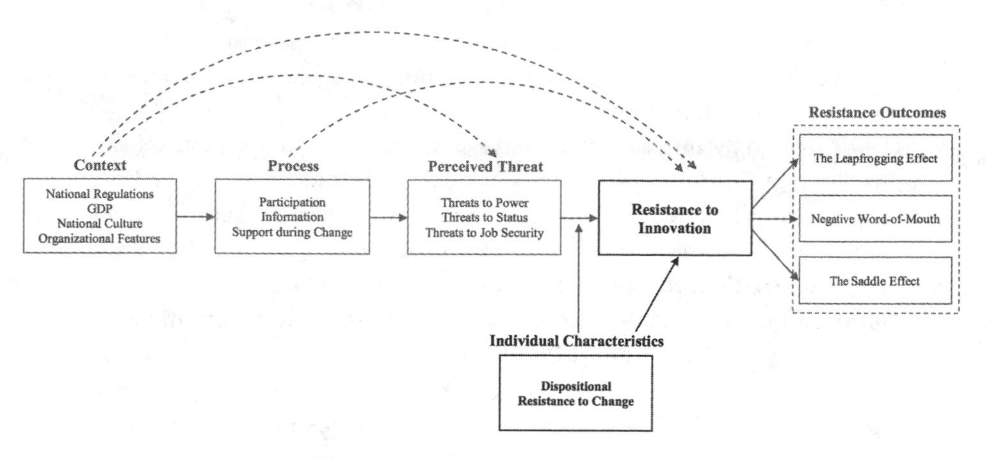

FIGURE 1.1. Individuals' characteristics and resistance to innovation.

to their predetermined schedule or change some of their courses. In one of the earlier studies in which the RTC scale was validated, the idea was to see whether the scale could be used to predict students' schedule change behaviors (Oreg 2003, study 5). Are students who dispositionally resist change less likely to make changes to their schedules? Following the add/drop period at the beginning of the academic school year, a sample of Cornell University students filled out the RTC scale and were asked to report whether they have made any changes to their course schedules. As predicted, the higher a student scored on the RTC scale, the less likely he/she was to have changed schedules. Furthermore, along with the RTC scale, a series of other, related scales were also administered, including measures of sensation seeking, tolerance for ambiguity, dogmatism, and several others. The RTC scale was the only scale to yield a significant effect on students' change behavior, thus establishing its incremental validity and conceptual distinctiveness.

In a following study (Oreg 2003, study 6), the focus was on university instructors and their willingness to adopt a new online course management system. This was "CourseInfo," which later became the Blackboard Learning Management System. The system provides instructors with a standardized means of communicating with students online. Codeveloped at Cornell University, Cornell was one of the first universities to implement CourseInfo in 1997 and offer faculty members the option of using it in their courses. Its initial diffusion was rather slow, with a rapid increase in its adoption rate commencing in 2002 (Fabian 2006). Yet our focus in this chapter is not on the overall adoption trend, but rather on explaining individual differences in adoption behavior. As with any product, some instructors preferred waiting longer than others before adopting the system. Could we predict who these instructors are?

In 2000, only three years after the first version of the system was introduced at Cornell, I (S. O.) conducted a small study to see if the RTC scale could predict adoption of the CourseInfo system. Instructors from a number of departments at Cornell were contacted via e-mail. Sixty-seven responded, only forty-seven of whom have been at Cornell since the first introduction of CourseInfo. Respondents were asked to fill out the RTC scale and to report whether they use the CourseInfo system, and for how long. It was predicted that the higher an instructor's RTC score, the less likely will he or she have adopted the system. This was indeed what I found. Furthermore, among those who adopted it, higher RTC scores were associated with later adoption.

Similar results were found with respect to consumers' adoption of electronic products (Oreg, Goldenberg, and Frankel 2005). In a study of faculty members in Israel, after filling out the RTC scale, participants were asked to report the approximate dates at which they purchased a variety of electronic products, such as their first computer and cell phone. As expected, higher RTC scores were associated with later adoption dates.

In a more recent set of studies, Oded Nov and Chen Ye (2009; 2008) used an abbreviated version of the RTC scale and showed that it indirectly predicted potential users' intentions to adopt digital libraries. Their findings suggest that the disinclination of high RTC individuals to using digital libraries comes from the perceived ease of use. The higher were their participants' RTC scores, the more difficult these participants believed it would be to use the digital library system.

Again, similar findings were obtained in a study of librarians and their willingness to learn about and use Web 2.0 applications, which are becoming central for the library and information science community (Aharony 2009). Higher RTC scores were associated with lower willingness to use Web 2.0 applications. Librarians with higher RTC scores also expressed lower motivations to learn about Web 2.0 and perceived them to be less important. Thus in addition to directly influencing individuals' adoption behavior, RTC may yield less adoption because of its influence on how innovations are perceived.

Turning to Rogers's (1962; 2003) description of the adoption process, it would appear that at least part of the impact of dispositional resistance to change on adoption intentions is through its impact on the persuasion and decision stages, whereby dispositional resistors are less likely to seek information about the innovation, more likely to perceive it as too complex, and therefore more likely to evaluate it negatively.

Conclusions

Our focus in this chapter was on the role of personality in explaining individuals' resistance to innovation. Knowing one's dispositional orientation toward change can help predict the likelihood that innovations in general will be resisted. Marketers, however, don't typically have access to customers' personality profiles. More importantly, products are not typically marketed to individuals, but to whole markets. And even if marketers could tell ahead of time what characteristics their customers have, it's not as if one could change these characteristics to match the type of innovation being marketed. What use, then, is there for knowledge about the link between consumer characteristics and resistance? Well, as with respect to the remaining chapters of this book, understanding the reasons for resistance can often help circumvent it, even when the source of resistance cannot be directly changed. Although companies can't change their consumers' attributes, they can, and often do, adapt their marketing strategies to accommodate them.

Findings on dispositional resistance to change tell us that some people are disinclined toward change and innovation because they like their routines; they feel uneasy and sometimes even threatened by the notion of change. They focus on the short-term inconveniences of change, and they tend to hold on to their opinions and a priori decisions. At least some of these reasons for resistance, primarily routine seeking and short-term focus, can be easily addressed. Marketers often emphasize the novelty in their products. When a new brand is launched, the word *new* will often be highlighted on its package. Yet we know from the RTC routine-seeking dimension that to those who are dispositionally resistant to change this novelty could actually be a turnoff. These individuals know what they like, and stick to it. For them, alongside the explanations for how a product is better than its predecessor, as they typically appear in ads, an emphasis on how the product or its use are similar to earlier products could alleviate some of the resistance.

Knowing that part of these individuals' resistance also comes from their focus on the short-term inconveniences of change can help design specific marketing strategies for alleviating resistance. The short-term inconvenience in adopting an innovation typically comes from having to

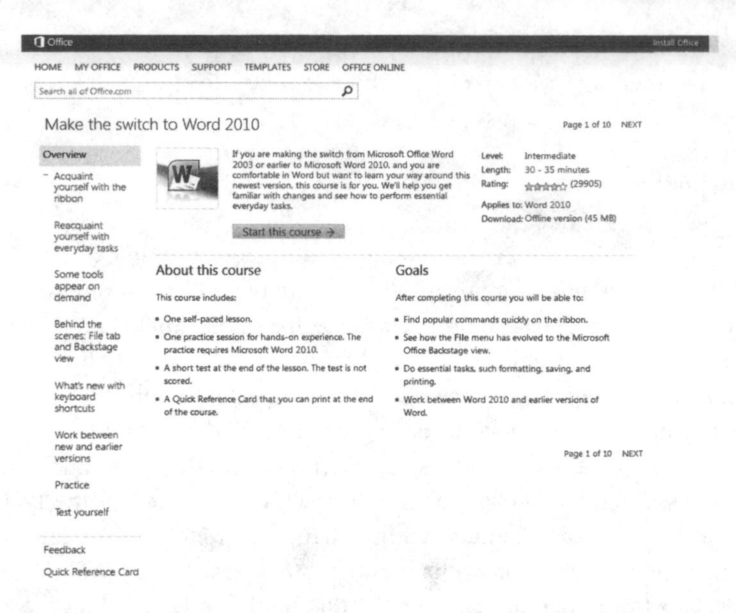

FIGURE 1.2. Screen shot of the "Make the Switch to Word 2010" web page. Downloaded from http://office.microsoft.com/en-us/word-help/make-the-switch-to-word-2010-RZ101816356.aspx.

adapt to new product features and to sometime regain proficiency in using the product. Marketers can address this source of resistance by offering support for the adjustment and learning period.

As we discuss in chapter 3, several companies do precisely this when launching product upgrades. For example, in many respects Microsoft's Office 2007 and 2010 programs could be considered radical in comparison to the 2003 versions because they employ an entirely different means of navigating through menus. Indeed, the new software was met with much resistance, primarily from experienced Office users who found the transition to the new "ribbon" confusing. Resistance is likely to have been highest among people with a short-term focus. Recognizing such resistance, Microsoft offered specialized tutorials tailored to those proficient with the 2003 versions, such as the "Guides to the Ribbon: Use Office 2003 Menus to Learn the Office 2007 User Interface," and the interactive "Word 2003 to Word 2007 Command Reference Guide." When Office 2010 was introduced, newer tutorials were offered for helping users "make the switch" to the 2010 programs (for example, see figure 1.2). These kinds of guides are designed to help consumers work through the short-term inconveniences that are involved in adopting the innovation, and are therefore likely to alleviate their resistance.

The focus of this chapter was on stable internal aspects, inherent to individuals, that predispose them to resist innovations, whatever these may be. Alongside such internal factors, the particular features of a given innovation will clearly contribute to the ultimate degree of resistance to it. As we elaborate in the following chapter, resistance to innovations often results from the threats that innovations may pose to potential adopters. Alongside their potential, and sometimes obvious, benefits, innovations can also threaten many aspects in individuals' lives. These threats may be less obvious to the observer, yet may be even more potent than the innovation's potential benefits.

Note

1. For comparison, analyses were also conducted using a Varimax rotation and using the Principle Axis extraction method with both Varimax and Promax rotations. Although factors were sometimes ordered differently, these analyses produced equivalent scale dimensionality.

References

Aharony, N. 2009. "Web 2.0 Use by Librarians." *Library and Information Science Research* 31 (1): 29–37.

Ashford, Susan J. 1988. "Individual Strategies for Coping with Stress during Organizational Transitions." *Journal of Applied Behavioral Science* 24 (1): 19–36.

Bartunek, Jean M., Catherine A. Lacey, and Diane R. Wood. 1992. "Social Cognition in Organizational Change: An Insider-Outsider Approach." *Journal of Applied Behavioral Science* 28 (2): 204–23.

Bartunek, Jean M., and Michael K. Moch. 1987. "First-Order, Second-Order, and Third-Order Change and Organization Development Interventions: A Cognitive Approach." *Journal of Applied Behavioral Science* 23 (4): 483–500.

Budner, Stanley. 1962. "Intolerance of Ambiguity as a Personality Variable." *Journal of Personality* 30: 29–50.

Coch, Lester, and John R. P. French Jr. 1948. "Overcoming Resistance to Change." *Human Relations* 1: 512–32.

Conner, Daryl. 1992. *Managing at the Speed of Change: How Resilient Managers Succeed and Prosper Where Others Fail.* 1st ed. New York: Villard Books.

DeVellis, R. F. 1991. *Scale Development: Theory and Applications.* Newbury Park, CA: Sage.

Digman, John M. 1990. "Personality Structure: Emergence of the Five-Factor Model." *Annual Review of Psychology* 41: 417–40.

Ellen, P. S., W. O. Bearden, and S. Sharma. 1991. "Resistance to Technological Innovations: An Examination of the Role of Self-Efficacy and Performance Satisfaction." *Journal of the Academy of Marketing Science* 19 (4): 297–307.

Fabian, C. A. 2006. "A Five-Year Analysis of University Initiatives to Increase Faculty Engagement." *Journal of Computing in Higher Education* 17 (2): 3–24.

Frenkel-Brunswik, E. 1948. "Tolerance toward Ambiguity as a Personality Variable." *American Psychologist* 3: 268.

Goldberg, Lewis R. 1982. "From Ace to Zombie: Some Explorations in the Language of Personality." In *Advances in Personality Assessment*, edited by C. D. Spielberger and J. N. Butcher. Vol. 1. Hillsdale, NJ: Lawrence Erlbaum Associates.

———. 1990. "An Alternative 'Description of Personality': The Big-Five Factor Structure." *Journal of Personality and Social Psychology* 59 (6): 1216–29.

———. 1994. "Basic Research on Personality Structure: Implications of the Emerging Consensus for Applications to Selection and Classification." In *Personnel Selection and Classification*, edited by Clinton B. Walker and Michael G. Rumsey. Hillsdale, NJ: Lawrence Erlbaum Associates.

Goldsmith, Ronald E. 1984. "Some Personality Correlates of Open Processing." *Journal of Psychology: Interdisciplinary and Applied* 116 (1): 59–66.

Goldsmith, Ronald E., Jon B. Freiden, and Jacqueline K. Eastman. 1995. "The Generality/Specificity Issue in Consumer Innovativeness Research." *Technovation* 15 (10): 601.

Hannan, M., and J. Freeman. 1984. "Structural Inertia and Organizational Change." *American Sociological Review* 49: 149–64.

Harrison, Albert A. 1968. "Response Competition, Frequency, Exploratory Behavior, and Liking." *Journal of Personality and Social Psychology* 9 (4): 363–68.

Harrison, Albert A., and Robert B. Zajonc. 1970. "The Effects of Frequency and Duration of Exposure on Response Competition and Affective Ratings." *Journal of Psychology* 75 (2): 163–69.

Hinkin, Timothy R. 1998. "A Brief Tutorial on the Development of Measures for Use in Survey Questionnaires." *Organizational Research Methods* 1 (1): 104–21.

Hirschman, E. C. 1980. "Innovativeness, Novelty Seeking, and Consumer Creativity." *Journal of Consumer Research* 7: 283–95.

Im, S., B. L. Bayus, and C. H. Mason. 2003. "An Empirical Study of Innate Consumer Innovativeness, Personal Characteristics, and New-Product Adoption Behavior." *Journal of the Academy of Marketing Science* 31 (1): 61–73.

Jacoby, Jacob. 1971. "Multiple-Indicant Approach for Studying New Product Adopters." *Journal of Applied Psychology* 55 (4): 384–88.

Judge, Timothy A., Carl J. Thoresen, Vladimir Pucik, and Theresa M. Welbourne. 1999. "Managerial Coping with Organizational Change: A Dispositional Perspective." *Journal of Applied Psychology* 84 (1): 107–22.

Kanter, R. M. 1985. "Managing the Human side of Change," *Management Review* 74: 52–56.

Kruglanski, Arie W. 1989. *Lay Epistemics and Human Knowledge: Cognitive and Motivational Bases*. New York: Plenum Press.

Lau, Chung-Ming, and Richard W. Woodman. 1995. "Understanding Organizational Change: A Schematic Perspective." *Academy of Management Journal* 38 (2): 537.

Lee, K., and M. C. Ashton. 2004. "Psychometric Properties of the HEXACO Personality Inventory." *Multivariate Behavioral Research* 39 (2): 329–58.

Manning, Kenneth C., William O. Bearden, and Thomas J. Madden. 1995. "Consumer Innovativeness and the Adoption Process." *Journal of Consumer Psychology* 4 (4): 329–45.

Marcati, A., G. Guido, and A. M. Peluso. 2008. "The Role of SME Entrepreneurs' Innovativeness and Personality in the Adoption of Innovations." *Research Policy* 37 (9): 1579–90.

McCrae, Robert R., and Paul T. Costa. 1987. "Validation of the Five-Factor Model of Personality across Instruments and Observers." *Journal of Personality and Social Psychology* 52 (1): 81–90.

Mehrabian, A., and J. A. Russell. 1973. "A Measure of Arousal Seeking Tendency." *Environment and Behavior* 5 (3): 315–33.

Midgley, David F., and Grahame R. Dowling. 1978. "Innovativeness: The Concept and Its Measurement." *Journal of Consumer Research* 4 (4): 229–42.

———. 1993. "A Longitudinal Study of Product Form Innovation: The Interaction between Predispositions and Social Messages." *Journal of Consumer Research* 19 (4): 611–25.

Nov, Oded, and Chen Ye. 2008. "Users' Personality and Perceived Ease of Use of Digital Libraries: The Case for Resistance to Change." *Journal of the American Society for Information Science and Technology* 59 (5): 845–51.

———. 2009. "Resistance to Change and the Adoption of Digital Libraries: An Integrative Model." *Journal of the American Society for Information Science and Technology* 60 (8): 1702–08.

Oreg, Shaul. 2003. "Resistance to Change: Developing an Individual Differences Measure." *Journal of Applied Psychology* 88: 680–93.

———. 2009. "A Call for Greater Caution in Drawing Conclusions from Individual Samples: A Comment on 'A Test of the Measurement Validity of the Resistance to Change Scale in Russia and Ukraine.'" *Journal of Applied Behavioral Science* 45 (4): 490–93.

Oreg, Shaul, Mahmut Bayazit, Maria Vakola, Luis Arciniega, Achilles A. Armenakis, Rasa Barkauskiene, Nikos Bozionelos, Ivana Feric, Yuko Fujimoto, Luis Gonzalez, Jian Han, Hilde Hetland, Martina Hrebickova, Nerina L. Jimmieson, M. Kotrla, J. Kordacova, Hitoshi Mitsuhashi, Boris Mlacic, Sandra Ohly, P. Saksvik, I. Saksvik, and Karen van Dam. 2008. "Dispositional Resistance to Change: Measurement Equivalence and the Link to Personal Values across 17 Nations." *Journal of Applied Psychology* 93 (4): 935–44.

Oreg, Shaul, Jacob Goldenberg, and Rachel Frankel. 2005. "Dispositional Resistance to the Adoption of Innovations." Paper presented at the annual meeting of the European Association of Work and Organizational Psychology, Istanbul, Turkey.

Pally, S. 1955. "Cognitive Rigidity as a Function of Threat." *Journal of Personality* 23: 346–55.

Pearson, P. H. 1970. "Relationships between Global and Specified Measures of Novelty Seeking." *Journal of Consulting and Clinical Psychology* 34 (2): 199–204.

Raju, P. S. 1980. "Optimum Stimulation Level: Its Relationship to Personality, Demographics, and Exploratory Behavior." *Journal of Consumer Research* 7 (3): 272–82.

Rogers, Everett M. 1962. *Diffusion of Innovations*. New York: Free Press of Glencoe.

———. 1995. *Diffusion of Innovations*. 4th ed. New York: Free Press.

———. 2003. *Diffusion of Innovations*. 5th ed. New York: Free Press.

Rokeach, Milton. 1960. *The Open and Closed Mind*. New York: Basic Books.

Sagie, Abraham, and Meni Koslowsky. 2000. *Participation and Empowerment in Organizations: Modeling, Effectiveness, and Applications*. Thousand Oaks, CA: Sage.

Saksvik, Ingvlid Berg, and Hilde Hetland. 2009. "Exploring Dispositional Resistance to Change." *Journal of Leadership and Organizational Studies*. doi: 10.1177/1548051809335357.

Schwartz, Shalom. 1992. "Universals in the Content and Structure of Values: Theoretical Advances and Empirical Tests in 20 Countries." In *Advances in Experimental Social Psychology*. Vol. 25, edited by Mark P. Zanna. San Diego: Academic Press.

Slovic, P. 1972. "Information Processing, Situation Specificity, and the Generality of Risk Taking Behavior." *Journal of Personality and Social Psychology* 22: 128–34.

Stewart, Wayne H., Jr., Ruth C. May, Daniel J. McCarthy, and Sheila M. Puffer. 2009. "A Test of the Measurement Validity of the Resistance to Change Scale in Russia and Ukraine." *Journal of Applied Behavioral Science* 45 (4): 468–89.

Tichy, Noel M. 1983. *Managing Strategic Change: Technical, Political, and Cultural Dynamics*. New York: Wiley.

Venkatraman, Meera P. 1990. "Differentiating between Cognitive and Sensory Innovativeness: Concepts, Measurement, and Implications." *Journal of Business Research* 20 (4): 293–315.

Wanberg, Connie R., and Joseph T. Banas. 2000. "Predictors and Outcomes of Openness to Changes in a Reorganizing Workplace." Journal of Applied Psychology 85 (1): 132–42.

Watson, Goodwin. 1971. "Resistance to Change." *American Behavioral Scientist* 14 (5): 745–66.

Werner, H. 1946. "Abnormal and Subnormal Rigidity." *Journal of Abnormal and Social Psychology* 41 (1): 15–24.

Zaltman, Gerald, and R. Duncan. 1977. *Strategies for Planned Change*. New York: Wiley.

Zuckerman, Marvin. 1994. *Behavioral Expressions and Biosocial Bases of Sensation Seeking*. New York: Cambridge University Press.

Zuckerman, Marvin, and Kathryn Link. 1968. "Construct Validity for the Sensation-Seeking Scale." *Journal of Consulting and Clinical Psychology* 32 (4): 420–26.

CHAPTER TWO

What's in It for Me, and What Do I Have to Lose?

Practical Reasons for Resisting Innovation

One of the greatest and most fundamental discoveries in physics involves the notion that electricity, magnetism, and light are all manifestations of what is termed the *electromagnetic field*. This notion, known as Maxwell's theory, was introduced by James Maxwell through a series of papers published in the 1860s and 1870s (Maxwell 1865; 1861; 1873). Although it has now been widely accepted for more than a century, the theory was initially met with fierce resistance, and it took more than thirty years following its first introduction before its practical implications were readily utilized in the work of engineers (Hunt 1983). As we will describe, this famous struggle exemplifies one of the most salient factors to elicit resistance to innovations—the perception of threat.

One of the main antagonists of Maxwell's theory and its supporters was William Preece, chief electrician of the British Post Office, in charge of the country's telegraphs and telephones and a leading authority on electrical engineering (see figure 2.1). Preece's authority was based on his scientific prestige, as a fellow of the Royal Society, and on a long list of honors, including being knighted in 1899, after becoming president of the Society of Telegraph Engineers (in 1880 and again in 1893), of the British Association's engineering section (in 1888), and of the Institution of Civil Engineers (in 1898; Hunt 1983). Representing the camp of "practical men," wanting to preserve the simple formulas and procedures that had worked in the past, Preece saw Maxwell's theory as a threat.

While Maxwell's followers, including scientists Oliver Lodge, George FitzGerald, and Heinrich Hertz, and former telegrapher Oliver Heaviside,

FIGURE 2.1. William Preece, carrying the banner of "experience" on a lightning rod, striding over Maxwellian supporter Oliver Lodge, lying in the mud with his rod and "experiment" banner. The illustration, titled "Ajax Defying the Lightning," appeared on the cover of the December 1888 issue of *Electric Plant*.

were making significant strides in further establishing Maxwell's theory through their experiments, Preece adamantly disparaged their findings. This intercourse between the two camps became known as the war of "practice versus theory" (Hunt 1983), which constituted "an attempt by the practical men to hold their own ground against the incursions of the scientists. . . . The issue of authority lay at the heart of the conflict between practice and theory, and made it especially bitter" (Hunt 1983, 351–52).

Following a series of publications by the Maxwellian group, Preece had conducted his own (later to be proven flawed) experiments and published findings that allegedly undermined Maxwell's theory. When Heaviside tried to publish evidence disproving Preece's claims, Preece used his position at the post office, whose censors had to clear articles on telephony before they could be published, to block the article and prevent its publication. Soon after, other versions of Heaviside's work were similarly rejected from publication in other outlets. Harsh exchanges also transpired between Preece and the physicist Lodge. Demonstrating the degree of hostility that characterized the debate was an illustration published on the cover of the December 1888 issue of the *Electric Plant* in which Preece's image, representing "experience" is seen trampling over the image of Lodge, representing the "experiment." It was only a matter of time, however, for the accumulation of findings from the Maxwellian camp's experiments to overcome resistance, leading to the wholehearted acceptance of Maxwell's theory by even the "practical men."

Such resistance to new ideas is not uncommon among scientists. In fact, the system through which scientific discoveries are typically revealed—the peer-review system—is often the source of resistance to innovative ideas (Campanario 1996; Campanario 2002). Some have suggested that the peer-review process may be "adequate for the middle of the road, unadventurous article, but [is] hopeless for the one with new and challenging ideas" (Lock 1982, 1225). Implicating the role of power in this resistance, English biologist Thomas Henry Huxley, one of the key promoters of Darwin's theory of evolution, wrote the following in 1852 in a letter to his sister (published in Huxley and Huxley 1900):

> I have just finished a Memoir for the Royal Society, which has taken me a world of time, thought, and reading, and is, perhaps, the best thing I have done yet. . . . You have no notion of the intrigues that go on in this blessed world of science. Science is, I fear, no purer than any other region of human activity; though it should be. Merit alone is very little good; it must be backed by tact and knowledge of the world to do very much.
>
> For instance, I know that the paper I have just sent in is very original and of some importance, and I am equally sure that if it is referred to the judgment of my "particular friend," [Mr. X], that it will not be published. He won't be able to say a word against it, but he will pooh-pooh it to a dead certainty.
>
> You will ask with some wonderment, Why? Because for the last twenty years, [Mr. X] has been regarded as the great authority on these matters, and has had no one to tread on his heels, until at last, I think, he has come to look upon the Natural World as his special preserve, and "no poachers allowed." So I must maneuver a little to get my poor memoir kept out of his hands.

Both the Maxwell and Huxley cases demonstrate the role that power can have in the formation of resistance. Threats to one's position of authority and power, can be extremely powerful motivators to block new players and ideas from entering the playing field. We should distinguish, however, between resistance because of the experience of threat, and resistance because the new ideas simply appear to be incorrect. Although it is not uncommon for those who feel threatened to also genuinely believe that the new ideas are erroneous, the two reasons for resisting an innovation are nevertheless distinct.

Resisting new ideas because they are initially judged to be erroneous is a natural and healthy part of the scientific process. One such example involves Georges Lemaître's "Primeval Atom hypothesis," which became known as the Big Bang theory. Lemaître's work, based on Einstein's theory of relativity, suggested that the universe was expanding. In an encounter between Lemaître and Einstein shortly following Lemaître's publication, Einstein commented to Lemaître: "Your calculations are correct, but your physics is abominable" (cited in Singh 2004, 160). Such rejection from Einstein meant rejection by the establishment. In his book about the Big Bang theory, Singh writes: "In the absence of hard evidence, Einstein's blessing or criticism had the power to make or break a nascent theory. Einstein, who had once been the epitome of rebellion, had become an unwitting dictator" (160). Following further observational findings by Hubble, Einstein came to realize his mistake and publicly admitted it to Lemaître. Although Einstein's authority is what stymied Lemaître's discovery, resistance here does not seem to have resulted from any perceived threat by Einstein. If anything, Lemaître's theory supported rather than conflicted with Einstein's. Thus resistance in this case seemed to be substantive. Contrarily, our focus in the present chapter will be on those occasions, as in the cases opening this chapter, in which resistance is the product of perceived threat.

All of the cases above involve, in one way or another, scientific *paradigm shifts*, a term coined by Thomas Kuhn in *The Structure of Scientific Revolutions* (1962). The term was later adopted to more generally describe a mechanism of revolutionary change in one's basic assumptions and theories. Kuhn terms a paradigm what members of a (scientific) community share, believe, and agree upon. A scientist considers alternative theories or models, but a shift occurs only when a prevailing paradigm is replaced by a new, competing one.

Once a paradigm shift is completed, the scientist can no longer retain the previous paradigm given the compelling evidence against it. This shift usually occurs rapidly, but only after a long period of resistance through which the old paradigm was embraced. In 1900, Lord Kelvin famously stated, "There is nothing new to be discovered in physics now." Five years

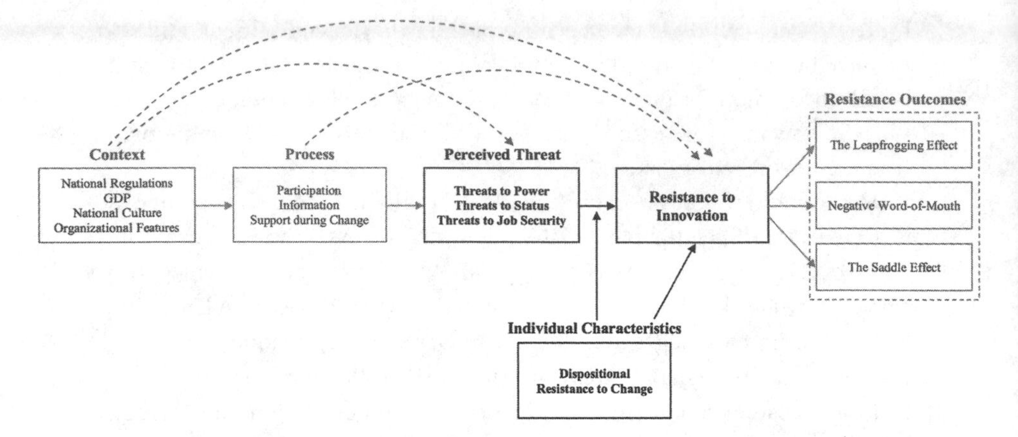

FIGURE 2.2. Perceived threat and resistance to innovation.

later, Einstein published his paper on special relativity, which, besides being one of the most creative scientific works to have been proposed, was a classic case of a paradigm shift from Newtonian mechanics, which had been used to describe force and motion for over two hundred years. As we elaborate below, we suggest that the resistance that precedes the paradigm shift will often result from threats that the potential adopter may perceive in the new paradigm. And although the examples we provided above are all within the scientific realm, resistance clearly emerges across realms and is often displayed toward new ideas, new technologies, or any other factor that serves to alter the status quo.

The impact of perceived threats on resistance is parallel to the role of perceived utility in explaining adoption (e.g., Rogers 2003). Instead of explaining that people adopt a product because they see value in it, one could just as well explain why people resist a product because they perceive it as posing a threat. Such a focus on resistance because of perceived threats makes the concept of resistance seem more rational than how it is typically viewed. Indeed, although the connotation of resistance to change is ordinarily negative, implying that something is wrong with the resistor, changes are often resisted for very sensible reasons. In the context of organizational change it has been suggested that what is resisted is not change per se, but the loss that change often entails (e.g., Dent and Goldberg 1999). In the organizational literature, this source of resistance is said to involve the *content* of the change rather than its *process*, as we discuss in the next chapter. Changes often bring about the potential loss in income, job security, working conditions, social ties, job autonomy, power, and expertise, to name a few. We focus in this chapter on these forms of loss and on their effect on individuals' resistance to innovations (see figure 2.2).

Sources of Threat

The concept of loss or threat is similar to the *risk barriers* Ram and Sheth (1989) had discussed, which included physical, economic, functional, and social barriers. By physical barriers Ram and Sheth referred to the threats of physical harm some innovations may present, such as in the case of new medication, or to name a timely product, cellular phones. In economic barriers they are referring to the high costs of some innovations. This definition of economic barrier could be extended to include cases in which the risk of economic loss comes not from the price of the innovation but rather from the negative implications that some innovations have on individuals' income. Ram and Sheth's functional barriers are about consumers' fears that the innovation may not perform in line with expectation. Finally, social barriers refer to the risk of being ostracized for purchasing certain products. Ram and Sheth provide the example of generic brands, but certainly many others could be suggested today, including products that are harmful to the environment.

Following Ram and Sheth's perspective, in the present chapter we discuss several other potential threats as sources of resistance to innovation, focusing in particular on the threat to one's expertise, which is highly relevant for the consumer context. We list some of the threats considered and the works in which they have been investigated in table 2.1.

Resistance Because of Threats to Job Security

In the organizational context, many innovations alter the way in which jobs are performed. Such alterations often involve some people losing their jobs. Clearly, the possibility of losing one's job is a major and obvious reason for resisting an innovation. As obvious as it is, however, managers and marketers are nevertheless often surprised when the innovations they introduce are met with resistance, and they typically underestimate the negative implications that threats to individuals' job security have on their well-being. In several studies of organizational change a key threat that is

TABLE 2.1 **Perceived threat categories and studies in which they have been investigated**

Domain of perceived threat	Relevant publications
Job security	(Ashford 1988)
	(Fried et al. 1996)
	(Gaertner 1989)
	(Naswall et al. 2005)
	(Oreg 2006)
	(Paulsen et al. 2005)
	(Vieitez et al. 2001)
Power and prestige	(Buhl 1974)
	(Goltz and Hietapelto 2002)
	(Lapointe and Rivard 2005)
	(Oreg 2006)
	(Tichy 1983)
	(Zaltman and Duncan 1977)
Distributive justice	(Armenakis et al. 2007)
	(Bernerth et al. 2007)
	(Fried et al. 1996)
	(Paterson and Cary 2002)
	(Shapiro and Kirkman 1999)
	(Spreitzer and Mishra 2002)
Expertise and attained knowledge	(Ashford 1988)
	(Lapointe and Rivard 2005)
	(Moreau et al. 2001)
Job factors (e.g., changes in demands and complexity)	(Axtell et al. 2002)
	(Bartunek et al. 2006)
	(Caldwell et al. 2004)
	(Fedor et al. 2006)
	(Fried et al. 1996)
	(Hall et al. 1978)
	(Herold et al. 2007)
	(Morse and Reimer 1956)
	(Oreg 2006)
	(Susskind et al. 1998)
	(van Dam 2005)

described as a source of resistance is the threat of job loss. In one study, a Swedish hospital was undergoing changes, with the aim of reducing costs and improving productivity, through the implementation of new managerial practices, closing or merging departments, and small-scale layoffs (Naswall, Sverke, and Hellgren 2005). Data from four hundred nurses indicated that above and beyond the effects of employees' personality (see chapter 1), the degree to which nurses experienced uncertainty and threat to their employment significantly predicted their reaction to the change, as manifested in their psychological mental health and experienced stress.

Similar findings were obtained in a study of an automobile manufacturing company, facing the introduction of new technology (Vieitez, De La

Torre Carcía, and Vega Rodríguez 2001). The company introduced a system of Advanced Manufacturing Technology (AMT), which involved the automation of production lines and a computerized management system for controlling productivity. A range of responses to the change emerged, with some employees expressing particularly high levels of anxiety and depression following the change. In line with the researchers' hypotheses, it was the experience of job insecurity following the introduction of the technology that elicited the anxiety and depression.

The notion that technological innovations are threatening individuals' job security is clearly and explicitly proposed in Jeremy Rifkin's *The End of Work* (1995). With examples from several industries Rifkin demonstrates the effects that technological innovations have had on employment rates. In the farming industry he provides the example of the Robotic Melon Picker (a.k.a. ROMPER), developed through collaborations of researchers from Israeli universities and Purdue University in Indiana. ROMPER is a tractor and picker that requires infrequent human assistance in guiding itself down rows of melon plants. It uses a camera and special software that allows it to identify the melons, and following the use of a chemical sniffer to determine ripeness, uses a robotic arm to pluck the ripe melons. In field tests the ROMPER successfully harvested 85% of the melons (Edan 1995). Introduction of the ROMPER, Rifkin writes, will have a dramatic effect on the thousands of Palestinians that Israeli farmers employ, many of whom will become unemployed.

In the automotive industry Rifkin describes other kinds of "smart" robots that are already replacing human labor. According to Rifkin, Japanese car manufacturers, through the use of automation, manage to produce the same amount of cars as US manufacturers with less than a fourth of the employees. In the near future, he suggested, machines with advanced capabilities, such as voice communication, coordination abilities, and self-navigating skills, will be used to further replace humans in manufacturing jobs. Even in the service industry technological innovations are replacing people at work. Voice recognition machines are replacing operators, Internet banking services are replacing human tellers, computerized hiring systems are replacing personnel employees in screening candidates, and machines with sight recognition are replacing postal workers in sorting mail. With such threats to people's jobs, it is no wonder that some innovations are being adamantly resisted. It is therefore bewildering that so many managers and marketers seem to disregard, or at the least underestimate, such negative implications of the innovations they introduce.

Resistance Because of Threats to Power

In the Maxwell case opening this chapter we provide an example of how innovations may threaten individuals' power and status in their field. In the context of organizational changes, threats to power have been highlighted in several studies as a key source of resistance (e.g., Goltz and Hietapelto 2002; Tichy 1983; Zaltman and Duncan 1977). Changes in organizational structures often entail employment shifts that increase the power of some individuals in the organization at the expense of others. Whereas those gaining power are inclined to support change, those who risk losing power will more naturally resist it.

With respect to power and resistance in the context of technological change, Lance Buhl describes and analyzes resistance to technological progress in the post–Civil War navy (Buhl 1974). The case pertained to the dismissal of the USS *Wampanoag* (see figure 2.3) in September 1869 by the Board on Steam Machinery. The board was chaired by Rear Admiral Louis Goldsborough, among the most respected senior line officers in the service. Despite the fact that the *Wampanoag* established the world's steam speed record, not to be met for over a decade to come by any other vessel, the board rejected it, arguing that it was "scarcely more than naval trash." On the face of it, the board's rejection was based on the fact that although fast, the *Wampanoag* was very inefficient. Its machinery was bulky, amounting to over a third of the vessel's weight, and to attain its speed, it consumed extremely large amounts of coal. As Buhl's analysis reveals, however, underneath this surface explanation of the board's decision lay a long-standing political struggle between line and staff (among whom were the engineers), known as the "line-staff controversy." The decision on the *Wampanoag* was not a singular event, but rather represented "a sweeping rejection of over half a century of steam engineering in the American navy" (Buhl 1974, 705). Indeed, although the navy pioneered in steam propulsion, commissioning its first steam-run vessel in 1814, forces of resistance had led it to all but abandon the development of steam engines for nearly thirty years thereafter. Whereas line officers represented the age of sail, the engineers saw themselves as the "heralds and makers of the new age" (725). As Buhl describes it:

> the technology . . . had become a pawn in an intense social struggle. Steam propulsion was for engineers the basis for their livelihood and status; for line officers, it constituted a threat. The former hailed technical innovation uncritically

FIGURE 2.3. "An Incident of the Late War with Great Britain . . . USS *Wampanoag* Escaping from the Channel Fleet after Destroying the Halifax Convoy, July Fourth, 1866"—Oil painting by John Charles Roach, 1984, depicting the fast cruiser *Wampanoag* performing her designed mission, in an imaginary conflict. US Naval History and Heritage Command photo courtesy of Charles R. Haberlein Jr. Disclaimer: Use of released US Navy imagery does not constitute product or organizational endorsement of any kind by the US Navy.

> because it was their ticket to enhanced status, esteem, and authority in the navy. The latter tended to denigrate technical advances in order to maintain a monopoly on power and its perquisites. It was unfortunate but inevitable that the technical issues became even more muddled as a result. (726)

Resistance Because of Threats to Self-Concept and Identity

Beyond threats to concrete and explicit aspects of our lives, such as our jobs or how much influence we have on others, ideas and innovations are also resisted because of their potential impact on how we perceive ourselves. In her research on brand relationship theory, Susan Fournier talks about the relationships that consumers develop with the brands of the products they buy (Fournier 1998; Fournier 2009). Fournier argues that people are not buying brands simply because they like them or because they work well, but rather, because they benefit from the meanings these brands provide. Several of these meanings are not quite tangible and have more to do with how people perceive themselves and their position in society. An old Jewish joke nicely exemplifies this notion: A man, stranded for years on a desert island, was finally rescued. His rescuers discover that he has built not one synagogue, but two. To their question, "why," he answers "Well, this one I pray in every day, but that other one, you wouldn't catch me stepping foot in it." In many respects, our identity is maintained not only by what we adopt, but just as much by what we resist.

A key case that Fournier provides for demonstrating the role that such identity-related meanings could have on consumers' resistance is that of "New Coke," introduced back in April 1985. In the midst of the Coca-Cola–Pepsi battles, and following the introduction of Diet Coke, it became apparent that the American public was interested in sweeter soft drinks. The company established "Project Kansas," comprising a secret group of specially selected executives, headed by the company's marketing vice president, for the purpose of developing a new taste that would beat the "Pepsi Challenge" (McCampbell 1997). Taste tests conducted as part of the project indicated that a large majority of consumers preferred a sweeter-flavored cola version (Fournier 2001; Oliver 1986). Only 10% to 12% expressed a clear preference for the standing version of the drink and indicated that they would be upset by a change. Encouraged by these results, the company further developed the new drink's flavor and ultimately launched "New Coke" on April 23, 1985.

As it turned out, the resistant 12% had a much more substantial impact on the product's success than the company had expected. It was this particularly loyal sector of consumers that, through their negative word of mouth (see chapter 6), led to the reinstatement of the old Coke formula, marketed as "Coca-Cola Classic." Coca-Cola's president, Donald Keough, admitted years later that "the simple fact is that all the time and money and skill poured into consumer research on the new Coca-Cola could not measure or reveal the deep and abiding emotional attachment to original Coca-Cola felt by so many people" (Oliver 1986).

It is this attachment that Fournier highlights as a key source of consumers' resistance to innovation. The relationship that consumers have with their brands often serves to reinforce their self-concepts, through mechanisms of self-worth and self-esteem (Fournier 1998). Changes in a brand to which a consumer is strongly attached can therefore threaten his or her self-concept. Accordingly, as a lesson from cases such as that of New Coke, Fournier suggests that company executives try to explicitly assess their products' brand quality relationship. This brand quality relationship involves several psychological aspects, including brand loyalty, interdependence (that is, the degree to which a brand is ingrained in consumers' daily life), consumers' passion for the product, and the brand's self-concept connection (that is, "the degree to which the brand delivers on concerns, tasks, or themes important to a person's identity" [Fournier 1998, 364]). By taking such psychological factors into consideration, company executives can better prepare for the design and launch of new products to avoid or overcome consumer resistances.

In other studies, researchers highlight the role of social identity in its impact on individuals' adoption behavior. For example, in a recent set of studies by Jonah Berger and Chip Heath (2008), they demonstrate how product adoption by one group can lead to product rejection by another. Very much like in the two-synagogues joke, Berger and Heath found in a field study that undergraduates stopped wearing a certain type of wristband when members of another group of undergraduates, perceived as "geeks," started wearing it. This result complements findings in other studies of identity signaling by demonstrating how in contrast to social factors such as conformity and imitation, which expedite the diffusion of innovation, other social factors, such as the process of divergence, can serve to prevent diffusion. Berger and Heath's rationale resonates with William Swann's work on self-verification and self-coherence (e.g., Swann, Rentfrow, and Guinn 2003) and is based on the assumption that what drives people's divergence is their need to ensure that others correctly recognize their identity. When innovations threaten this identity, resistance can be expected.

Empirical Studies of Combined Threats

Most typically, the threats elicited by change and innovation pertain to more than a single domain. Accordingly, in a number of studies, a mixture of potential losses or threats was considered as the source of resistance. For example, in a study of an organizational restructuring, employees' resistance to the change was predicted by the degree to which they perceived a threat to their job security, power and prestige, and the degree of challenge, stimulation, and autonomy they had in their jobs (Oreg 2006, see additional information about this study below and in chapter 3). In line with previous studies, the perceived threat to job security was associated with negative emotional responses to the change, including increased stress and anxiety. Similarly, perceived threats to power and prestige were associated with negative cognitive evaluations of the change and, perceived threats to employees' sense of autonomy and challenge on the job were associated with both emotional and cognitive resistance to the change.

In another study, of the Bell Telephone Systems divestiture, the degree of stress experienced during the change was associated with employees' fear of job loss, potential pay cuts, and the chance that their skills will become obsolete (Ashford 1988). Similar conclusions about the link between perceived loss and resistance to change have been drawn in a case Liette Lapointe and Suzanne Rivard (2005) describe, pertaining to the introduction of an Electrical Medical Records (EMR) system in a teaching hospital. EMR systems provide access to patients' records from different locations and replace the previously used paper files. In the hospital described, the particular EMR system that was implemented was selected following a thorough review of available systems by a multidisciplinary committee, including physicians, nurses, and other professionals. Following the system's implementation, however, physicians discovered that finding information on the system, such as lab test results or notes from earlier consultations, took them longer than finding information with paper files did. Some physicians reported this taking them an extra 1.5 to 2 hours per day. Having received their medical training with the old paper file system, the new system posed a threat to their previously established expertise. Furthermore, considering that physicians at this hospital were compensated on a fee-for-service basis, the extra time spent on finding information on the system meant a substantial financial loss. In addition, whereas prior to the implementation of the system physicians could prescribe medications and other treatments by verbally asking the nurses to handle them, the new system required that physicians make prescriptions directly through the system.

In response, several physicians expressed what they viewed as a threat to their status and power. It is therefore not surprising that resistance to the new system mounted, with conflicts emerging between physicians and nurses and later between physicians and the hospital's administration. These conflicts culminated in the resignation of several physicians and ultimately the dismissal of the hospital's CEO by the Department of Health. Resistance to the innovation was a direct result of the perceived threats that adopters experienced, in this case, threats to economic well-being, status, and power, as well as to their expertise. In the following section we elaborate on the role of expertise in explaining resistance through a description of several field studies we had conducted.

Resistance Because of Threats to Expertise and Attained Skills and Knowledge

In the late 1990s, an elite unit in the Israel Defense Forces decided to replace servicemen's personal firearms. The change had major implications on the unit and its members; besides the substantial cost of replacing a large amount of guns, such a change would also involve a great investment in training the entire force through the use of the new weapon. Clearly, it was an important decision, typically taken only once in twenty years. Through interviews with current and former unit members we learned that in preparation for choosing the new gun, a group of experts, including expert marksmen and arms specialists, was assembled to consider models from four manufacturers: (1) FN, (2) Jericho, (3) CZ, and (4) Glock.

The experts ranked the Glock, a new entrant to the guns market, as their least preferred option and recommended excluding it from further consideration. The main reasons were that the Glock appeared to be very different from anything they had handled before. With plastic components, it appeared more like a toy than a real, heavy-duty, weapon. Its safety mechanisms were very different from extant mechanisms, which would require them to undertake special training and adjustments. In addition, the experts complained that the shape of the gun's butt was substantially different than current models and would require them to adopt a different firing posture and to abandon their old techniques and habits. Furthermore, the manufacturer lacked the reputation of its competitors, which set further doubts in their minds. The experts' first choice was the latest FN pistol, which was an upgraded version of the FN guns already in use by the unit. A senior decision committee adopted the experts' recommendation and only the first three models were selected for the next selection stage, where more extensive testing was to be conducted.

These further tests were designed to simulate highly demanding tasks and extensive use through a broad range of combat scenarios. There is no clear record of why, but as it turned out, despite the early decision, the Glock found its way back in, to be included in these tests. According to one account, one of the senior commanders from the decision-making team, though not a shooting instructor himself, had a gut feeling during his morning jog that a mistake had been made in the early selection stage, and so he decided to include all four guns in the second stage of testing.

During the second testing phase, guns were subjected to long and extensive shooting periods to test both their performance and failure rates.

Much to the surprise of the experienced marksmen, the Glock emerged as superior to its three competitors in most of the parameters tested. By the time a decision was to be made, the Glock's advantages became clear to all, and it was chosen to replace the old models.

In hindsight, it appears that the right call was made and that the gun chosen did indeed fit the needs of the unit. The only reason, however, the Glock entered the testing phase in the first place was the spur-of-the-moment hunch of one commander, outside of the immediate expert group, who was less resistant and more willing to consider the item despite its unique attributes. Were it not for this, the Glock would not have been adopted.

Another example involves the introduction of the new official basketball in June 2006, to replace its predecessor of thirty-five years in the NBA. The new ball was said to include a new design and to be made of a new material that together offered a better grip, feel, and consistency than the previous leather ball. Players' resistance emerged immediately. Shaquille O'Neal, at the time a Miami Heat center, commented: "I think the new ball is terrible. . . . It's the worst decision some expert, whoever did it, made. . . . The NBA's been around how long? A hundred years? Fifty years? So to change it now, whoever that person is needs his college degree revoked. . . . Whoever did that needs to be fired. It was terrible, a terrible decision. Awful. I might get fined for saying that, but so what" (ESPN, October 3, 2006). Similarly, Miami Heat guard Dwyane Wade commented that "[this] will require another adjustment period. Now I've got to make another adjustment with a ball that I haven't shot with at all and it's going to be a challenge. . . . That means it's going to take a lot of late nights for me" (ESPN, October 3, 2006). Ultimately, resistance to the new ball was so strong that the NBA decided to shelve it and switched back to the old leather ball, beginning January 1, 2007. As NBA commissioner David Stern noted, "Although testing performed by Spalding and the NBA demonstrated that the new composite basketball was more consistent than leather and statistically there has been an improvement in shooting, scoring and ball-related turnovers, the most important statistic is the view of our players" (ESPN, December 12, 2006).

The examples above are of experts who resisted innovations. This may be surprising given that individuals often become experts by continuously examining and tinkering with new ideas. Experts make sure to keep up to date. Yet as we will explain, the link between expertise and innovation adoption depends on the type of innovation at hand. A clear distinction needs to be made between incremental innovations and radical ones. Whereas experts may be keen on adopting incremental innovations, the examples above suggest that they may resist radical ones. As we explain below, and distinct from cases in which innovations are resisted because of threats to power, experts' resistance is often a result of the threat to the relevance of their attained skills and knowledge (Oreg and Goldenberg 2007). Certainly many of the individuals who resist innovations because of threats to their positions of power, including William Preece and Louis Goldsborough in our examples above, are considered experts. Yet the two sources of resistance—threats to power and threats to expertise—are nevertheless distinct, as is the manner through which they influence resistance. Our focus in this section is on resistance that is due to experts' reluctance to give up their attained skills and knowledge and their desire to avoid having to relearn and regain their expertise. It is not about losing prestige, but rather about losing one's personal sense of mastery. Given the central role of expertise for the adoption of innovations in the consumer context, we elaborate on this potential source of threat and describe a number of studies we conducted to explore the interaction between expertise and innovation type in their influence on consumers' adoption intentions.

EXPERTISE AND RESISTANCE TO INNOVATIONS. The ability and willingness to adopt new ideas is largely a function of prior related knowledge (Cohen and Levinthal 1990). Through their discussion of organizations' *absorptive capacity*—"the ability . . . to recognize the value of new, external information, assimilate it, and apply it to commercial ends" (128), Wesley Cohen and Daniel Levinthal (1990) emphasize the role of related knowledge and expertise as a prerequisite for firms' innovativeness. More specifically, firms' absorptive capacity is said to rely on the expertise of gatekeepers, who reside at the interface between a unit and external sources of information. Their expertise puts such key members in a better position for identifying nascent ideas and enables them to help other members in the organization better understand the potential benefits of such ideas.

Outside of organizations, the equivalent of these experts are opinion leaders—individuals from whom others seek advice and information (Rogers and Cartano 1962). Opinion leaders within a given field are typically more knowledgeable about, and enduringly involved with, the relevant product class (e.g., Richins and Root-Shaffer 1988; Venkatraman

1990). Myers and Robertson (1972) examined the "knowledgeability" of opinion leaders in twelve categories using four hundred households in the Los Angeles area. As evidence for the expertise of opinion leaders, several product-related attributes have been found to correlate with opinion leadership, including *involvement and interest* (e.g., Coulter, Feick, and Price 2002; Myers and Robertson 1972; Richins and Root-Shaffer 1988; Summers 1970; Venkatraman 1990), *knowledge* (e.g., Coulter et al. 2002; Flynn et al. 1994; Flynn et al. 1996; Myers and Robertson 1972; Summers 1970; Venkatraman 1990), *usage* (e.g., Coulter et al. 2002), *awareness* (Coulter, Feick, and Price 2002; Goldsmith and Desborde 1991), *product ownership* (Childers 1986), and *confidence in choices* (Coulter, Feick, and Price 2002), with correlations ranging from a low of 0.37 (with interest in household furnishing) to a high of 0.87 (with knowledge about cosmetics and personal care). Opinion leaders can thus be considered expert consumers.

Experts are often defined as those who possess "both knowledge and experience in applying knowledge to a variety of problems within a domain" (Hinds, Patterson, and Pfeffer 2001, 1233). Novices may have only a limited amount of experience in a domain and are less proficient in performing tasks within it. Comparisons of experts and novices across a variety of domains demonstrated that when experts are faced with a task *within their domain of expertise*, they tend to automatically retrieve a solution method (Ericsson and Smith 1991). Contrarily, novices need to cognitively construct a representation of the task and need to devise a step-by-step solution.

Furthermore, experts' knowledge in their domain is qualitatively more accessible in comparison to novices (e.g., Johnson et al. 1981). Their memory and response times for problems within the domain are far superior to those of novices, and experts' processing of problems within their domain may become virtually automatic (e.g., Reingold et al. 2001). Knowledge is organized with a great degree of connectedness and cross-referencing of concepts, forming a cohesive structure (Bedard and Chi 1992). New information, *within the problem domain*, is quickly and effectively integrated with the existing body of knowledge (Patel and Groen 1986).

As the cases described above, however, suggest, experts' approach to innovations, even effective ones, will not always be favorable. On the contrary, many innovations may actually threaten individuals' expert position. Once individuals have mastered a given domain to the point of becoming experts, they will be reluctant to embrace innovations that, in essence, force them to forgo prior learning and expertise. Indeed, research on switching costs addresses the negative effect of early set-up costs (such as those incurred by experts) on individuals' propensity to adopt new technologies (e.g., Zauberman 2003).

In fact, a recurring albeit not emphasized factor in the discussion of absorptive capacity indicates that prior knowledge is an essential component in assisting the diffusion of innovations in organizations, as long as this prior knowledge is directly relevant (Cohen and Levinthal 1990). To assimilate new information, at least part of experts' prior knowledge needs to be closely related to the new knowledge. Furthermore, Cohen and Levinthal suggest that the ease and pace at which innovations are likely to be adopted is a function of "the degree to which an innovation is related to the pre-existing knowledge base of the prospective users" (Cohen and Levinthal 1990, 148). They go on and use the term *lockout* to describe situations in which new ideas are ignored or resisted because they are too dissimilar to extant knowledge. Indeed, whereas experts in the guns case above were eager to adopt the FN, which was very familiar to them, they resisted the Glock, which was radically different from what they had become accustomed to. Similarly, the NBA stars rejected the new game ball despite its superior objective performance. In all of these cases, concepts were underappreciated because they were qualitatively different from the current concepts in use.

Studies of the *transferability* of expertise across domains (e.g., Sims and Mayer 2002) suggest that transferability is a function of the extent to which the new domain shares rules with the current domain of expertise. As long as many rules are shared, experts will demonstrate their superior performance in new domains (Patel, Evans, and Groen 1989). However, when only few rules are shared, experts' performance will be equivalent, *or even inferior*, to that of novices (Bedard and Chi 1992). Indeed, presenting a problem for which one's expertise is not relevant may yield responses *that are even worse than those of novices*. For example, when presented with randomized chess positions (a situation entirely foreign to actual chess games), experts' recall of the chess position was worse than that of novice players (Chase and Simon 1973). This is because expert players organize information in meaningful chunks (for example, the Sicilian defense consists of one item to remember instead of sixteen positions). When information is categorized in large chunks (that only expert players master), it is easier to remember chess positions. When random positions are used, such chunks no longer exist and experts can no longer use their expertise. They may try to force a random situation to fit a familiar chunk, but this is often inappropriate and therefore ineffective.

Peter Frensch and Robert Sternberg (1989) suggest that the large size of experts' knowledge base sets them at a disadvantage in comparison to novices when trying to integrate new, incompatible information. Moreover, when new information is incompatible with existing knowledge, the increased proceduralization of strategies further increases inflexibility by limiting the knowledge selected and by reducing the level of control over the problem-solving process and hindering performance in new domains (e.g., Hesketh 1997; Marchant et al. 1991).

To maintain their expert position, experts work to extend their knowledge and deepen their specialization. When new concepts emerge, however, which are beyond experts' proceduralized set of habits, something changes in their set of considerations. The Glock, in the guns example above, was simply too different from extant guns; the gun's look and feel, its safety mechanism, butt, and weight, were all so different that the gun was considered inferior to the current model. Similarly, the new NBA ball was too different from its predecessor. Thus the relationship between individuals' level of expertise and their ability and motivation to adopt new ideas is a function of how similar the new ideas are to existing knowledge. It is therefore necessary to distinguish between innovations that are related to existing knowledge and those that are not.

INCREMENTAL VERSUS RADICAL INNOVATIONS. Innovations are often classified into incremental versus radical (the latter are also known as discontinuous, revolutionary, or breakthrough innovations [e.g., Dewar and Dutton 1986; Tushman and Anderson 1986]). Similar to paradigm shifts in science, radical innovations involve a departure from existing technology that is "so significant that no increase in scale, efficiency, or design can make older technologies competitive with the new technology" (Tushman and Anderson 1986, 441). An important difference between incremental and radical innovations involves the extent to which the innovation enhances or destroys previous competencies (Tushman and Anderson 1986). Whereas incremental innovations are based on previous technologies and can therefore improve performance through existing competencies, radical innovations render previous competencies useless because they introduce an entirely new approach to the problem at hand (see also Goldenberg et al. 2010). Thus whereas individuals' expertise will remain an asset when adopting incremental innovations, it may become a liability when adopting radical ones.

Contrarily, novices would not likely adopt incremental innovation for at least two reasons. First, they might not appreciate the value of the incremental innovation vis-à-vis the current technology they are using. Second, even if they do appreciate the differences between current technologies and the innovation, given their low level of expertise, the transition to the innovation, albeit being incremental, may seem too difficult to justify the effort. These two factors are underscored in the highly cited Technology Acceptance Model (TAM, Bagozzi, Davis, and Warshaw 1992; Davis, Bagozzi, and Warshaw 1989). Although not explicitly linked to the notion of expertise, TAM entails that individuals' intentions to use new technology will be a function of the technology's perceived usefulness and perceived ease of use. Whereas experts will be the first to appreciate the advantages of an incremental innovation, and will also be likely to perceive it as easy to use, this is not the likely case for novices.

We therefore argue that individuals' resistance to an innovation will be a function of the interaction between individuals' expertise and the type of innovation. More specifically, whereas novices' resistance to an innovation will be higher than experts' for incremental innovations, it will be lower for radical innovations. As we explained, experts resist radical innovations because these require them to relinquish their expertise with the previous technology. Contrarily, novices resist incremental innovations because they don't recognize the added value in adopting them over proceeding with current technologies.

A similar conceptual framework was tested in a set of studies by Page Moreau, Donald Lehmann, and Arthur Markman (2001). Moreau and colleagues' emphasis was on the degree to which experts and novices comprehended and saw the advantages in continuous (that is, incremental) and discontinuous (that is, radical) innovations. They suggested that whereas experts readily comprehend and see the advantages in continuous innovations, they fail to do so for discontinuous innovations. In contrast, novices were hypothesized to more readily comprehend and perceive the advantages of discontinuous innovations, but not continuous ones. Their framework was supported in two studies. In the first, the continuous product was a film-based camera with enhanced flash technology and the discontinuous product was a digital camera. In the second study, the continuous innovation was a "traditional" car with enhanced engine power and the discontinuous innovation was an electric car. Study participants who obtained high scores on knowledge tests relating to the tested products (cameras and cars) were considered experts. As predicted, experts better comprehended the continuous innovation and saw greater advantages in it; novices better comprehended and saw greater advantages in the discontinuous innovations.

Whereas the above framework closely corresponds with our own, it places its emphases on different aspects of the innovation. A key factor in our framework is on the degree to which the innovation threatens one's attained skill and knowledge base. Although Moreau and colleagues talk about the balance between the advantages versus risks that an innovation presents, they consider only risks that may hinder obtaining positive product outcomes (good pictures, a good driving experience). They do not consider, however, risks, or threats, to one's personal sense of mastery, which is key in our conceptualizations. Accordingly, whereas their assessment of expertise focused on the degree to which users possessed objective knowledge in the products' domain, our focus is on users' product-related experience. Our basis for selecting innovations for our own studies was therefore the degree to which proficiency in using previous technologies presented personal benefits for using incremental innovations, yet personal threats for using radical ones.

Study 1—Expertise and the Battle Between QWERTY and Dvorak

The innovations on which we focused in this study were computer keyboards. Accordingly, our sample included expert and novice typists. Experts were sampled from advanced secretarial courses and computer programming courses, who reported high typing rates, formal training in typing, and/or extensive typing experience. The final pool included sixty experts, consisting of computer programmers, secretaries, professional typists, and web designers. Approximately 60% of those approached agreed to participate in the study. Novices comprised first-year undergraduate students who reported having no substantial typing experience and considered themselves to be novice typists. Fifty-eight novices were recruited for the study. Seventy-five percent of those approached agreed to participate in the study. Among experts, 90% were women and the mean age was thirty-one (SD = 8.6). Among novices, 78% were women and the mean age was twenty-six (SD = 7.2).

We used two types of keyboards to represent incremental versus radical innovations. The standard and most prevalent keyboard used is the QWERTY, named after the key arrangement of the top-left-hand row of

FIGURE 2.4. The products introduced in study 1. The top panel portrays an outline of a plain Dvorak keyboard. The bottom panel portrays an ergonomic QWERTY keyboard. Sources: (*top*) Matias Corporation; (*bottom*) Hustvedt, from http://commons.wikimedia.org/wiki/File:QWERTY_Truly_Egronomic_Keyboard.jpg, no changes made.

keys. This arrangement originated in the need to reduce key jamming in old typewriters. It was not designed with typing speed as its main purpose, yet it remains the most common arrangement of most English keyboards today. We chose ergonomic keyboards that retain the QWERTY key arrangement to represent the incremental innovation (bottom of figure 2.4). These keyboards offer greater comfort than standard keyboards yet maintain the basic QWERTY arrangement.

Other key arrangements, however, have been proposed with typing speed as their primary goal. We chose one such keyboard, the Dvorak keyboard, named after its inventor, August Dvorak, as the radical innovation in our study (top of figure 2.4). The Dvorak keyboard was designed to minimize the distances that fingers are required to move across the keyboard. It is claimed to increase typing speeds by up to 40%, compared with QWERTY keyboards (David 1985).[1] Given that new key arrangements, such as those in Dvorak keyboards, render expertise on QWERTY keyboards practically useless, the Dvorak keyboard clearly constitutes a radical innovation.

To support our proposition that the Dvorak constitutes a radical innovation and the ergonomic QWERTY an incremental one, we ran a pilot study in which respondents were provided with definitions of incremental and radical innovations and were asked to rate the extent to which each of the keyboards should be considered incremental versus radical. The order in which keyboards were presented was counterbalanced. As expected, the Dvorak was rated as significantly more radical than the ergonomic QWERTY keyboard (t [35] = −5.30, $p < 0.001$). On a scale of 1 (entirely incremental) to 10 (entirely radical), the Dvorak keyboard received a mean rating of 6.5, whereas the ergonomic QWERTY received a mean rating of 4.0.

MEASURES AND PROCEDURE. Participants were asked to fill out questionnaires about their demographics (such as age, gender, level of education), their typing expertise (such as information about any formal training in typing they may have had, their typing speed, and such), and their intention to adopt each of the two products (based on product descriptions that were provided). Intention to adopt was evaluated by asking respondents about the attractiveness of each product and the likelihood that it would be purchased.[2]

RESULTS AND DISCUSSION. The difference in the amount of typing experience across the two subject groups was significant (t [114] = 7.50, $p < 0.001$), with an average of 11.27 years typing experience for experts and 3.19 for novices. Fifty-one percent of experts reported receiving formal training in touch typing, compared with only 8% of novices reporting such training.

To test our hypothesis, a Mixed Between (experts versus novices)-Within (incremental versus radical)-Subjects ANOVA was conducted, using intentions to adopt as the dependent variable. As expected, the expertise by innovation-type interaction term was significant (F [1115] = 18.221, $p < 0.001$). To interpret this finding, we plotted the interaction effect (figure 2.5). As figure 2.5 shows, whereas experts found incremental innovations more attractive than did novices, they found radical innovations significantly less attractive. To control for possible between-group

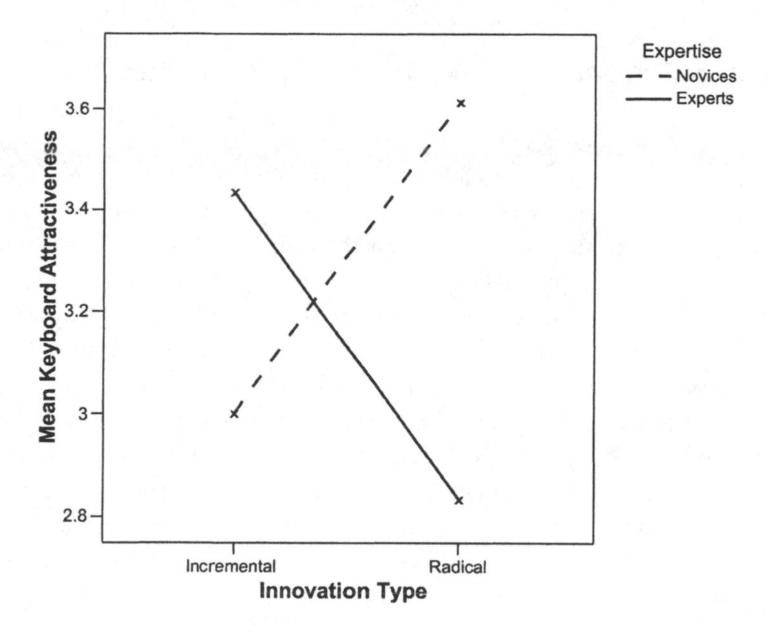

FIGURE 2.5. Expertise–Innovation type interaction in predicting keyboard attractiveness.

confounds, we reran the analysis while controlling for age and education level. The expertise by innovation-type interaction term remained significant ($p < 0.001$), and none of the control variables were significant.

These results support our hypothesis that whereas experts are keen to adopt incremental innovation they are more reluctant than novices when it comes to radical innovations. By adopting the radical innovation in this case, typists are essentially relinquishing their expertise and are required to start over in establishing it. On the other hand, novices, who have nothing invested in their current typing ability, are able to start fresh with a new, improved, typing technique.

In the context of this particular study, however, one can posit an alternative explanation for our findings. The key benefit of the radical innovation in this case (the Dvorak keyboard) is faster typing. Although, rationally, this would appear to offer a clear advantage, some employees may also perceive it as a shortcoming. For professional typists, typing faster may imply an increased typing load. The issue of typing loads, however, would not be of concern to novices, who did not type for a living. To discount such an alternative interpretation we replicate our results in study 2, in which work and effort considerations are irrelevant.

Study 2—Expertise and Adoption in a Gaming Context

The alternative interpretation we offered to our findings in study 1—that experts may be reluctant to adopt innovations that could increase their workloads—explains individuals' choices on the basis of extrinsic motivation. Such explanations of behavior focus on people's desire to obtain rewards while avoiding punishments (Skinner 1954). On the other hand, when individuals act on the basis of intrinsic motivation, what they strive for is personal competence and a sense of mastery and expertise (Deci and Ryan 1985). Collecting data in a different context, in which intrinsic motivation would appear to be the dominant factor, will allow for an independent test of our hypothesis, while ruling out explanations that are based on extrinsic motivation. Gaming contexts involve, almost by definition, intrinsic motivations, whereby the activity itself is motivating (Deci and Ryan, 1985). Thus, in the present study we approached a video-gaming context and used video console controllers as the innovations of study.

Our sample consisted of eighty-five video gamers (individuals who own and use home video games). Participants were sampled from two sets of sources, likely to yield high variance in expertise. First, we posted invitations in a variety of gamer forums on the Internet (for example, Playzone, Vgame). These forums attract individuals who devote much of their time to video games and typically hold high levels of expertise in gaming. The ad asked participants to contribute a few minutes of their time by answering questions about new equipment for a new game console. Data from forty-three respondents were collected. Twenty-three respondents were between the ages of eighteen and twenty-three, eleven were between twenty-three and thirty, and nine were between thirty-one and forty. Data from forty-two additional respondents were collected at the research assistant's place of work and in video game equipment stores. The age of two of these respondents was between eighteen and twenty-three, nineteen were between the ages of twenty-three and thirty, and twenty-one were between thirty-one and forty.

Both samples consisted of only male respondents, which is typical in the video gaming context. The final classification to expert and novice groups was conducted on the basis of three questions ("how important is it for you to be skilled in the use of your current game controller?" "how fast and proficient are you when using your controller?" and "how often do you play?"). The three items exhibited high internal consistency (Cronbach alpha was 0.82), and were averaged to create a composite expertise score. A median split was performed to classify individuals into expert and novice categories.

The incremental innovation consisted of an upgraded joystick controller. This new joystick was described as including additional movement sensors in the handle that allow for greater flexibility and greater control of the game characters' movements. The radical innovation consisted of a glove controller (see figure 2.6), which offers three-dimensional control with sensitivity and flexibility in user movements. With the glove, instead of achieving movement on the screen by shifting small levers and pushing buttons, movement is achieved by moving one's hand and fingers in midair, without the need to manipulate any external apparatus. Although this controller is aimed for use with the exact same games as current controllers, the approach to using it is entirely distinct and remote from that to which participants have become accustomed, thus meeting the defining criterion for radical innovations.

As in study 1, we ran a pilot study to test our assumption that the glove controller is a radical innovation and that the upgraded joystick is an incremental one. As expected, the glove controller was rated as significantly more radical than the upgraded joystick (t [35] = −7.79, $p < 0.001$). On a scale of 1 (entirely incremental) to 10 (entirely radical), the glove

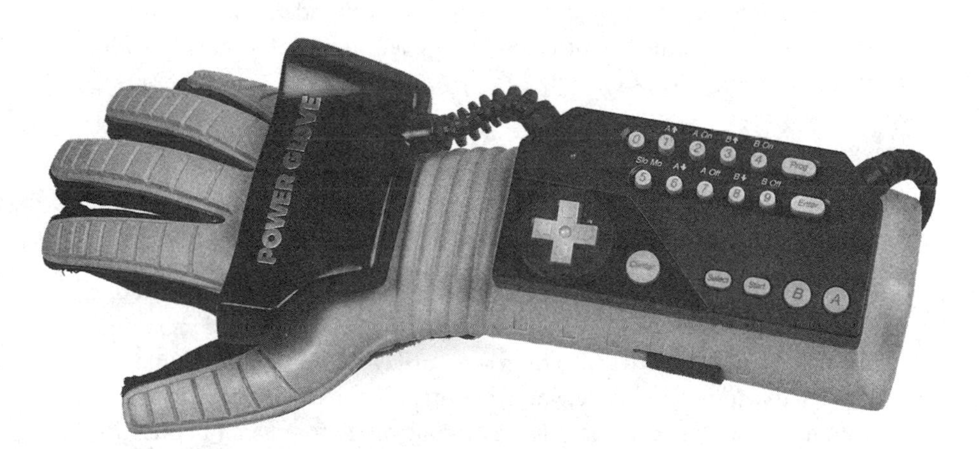

FIGURE 2.6. An American Power Glove controller for the NES, made by Mattel. Photographed by Evan Amos, 2011.

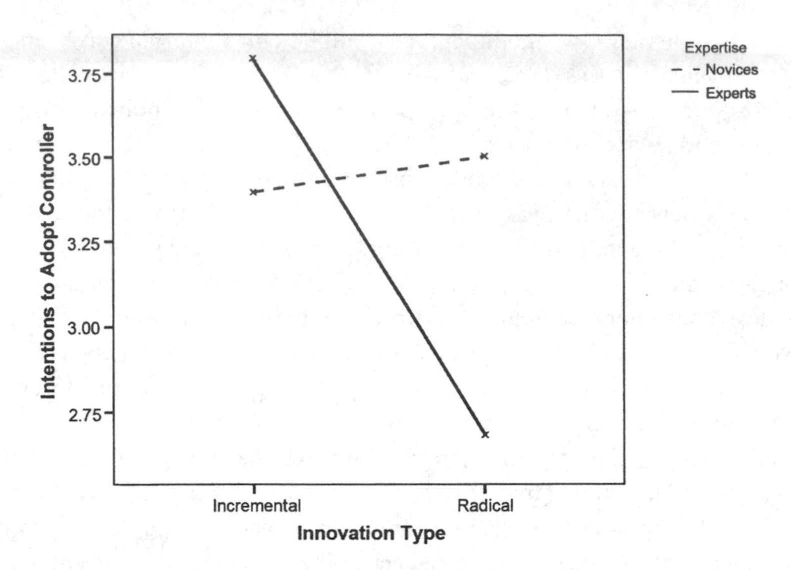

FIGURE 2.7. Interaction effect of expertise and innovation type on adoption intentions in study 2.

controller received a mean rating of 7.53, whereas the upgraded joystick received a mean rating of 4.17.

MEASURES AND PROCEDURE. Participants were randomly assigned a questionnaire with either the incremental or radical innovation. Each questionnaire consisted of questions about respondent demographics, their experience in the use of video games, and their intentions to adopt the innovation.

RESULTS AND DISCUSSION. To test our hypothesis, a 2 × 2 Between-Subjects ANOVA (experts versus novices and incremental versus radical) was conducted, using adoption intentions as the dependent variable. As expected, the expertise by innovation-type interaction term was significant (F [1, 84] = 14.1, $p < 0.001$). To interpret this finding, the interaction effect was plotted (figure 2.7). Similar to study 1, experts found incremental innovations more attractive than did novices but found radical innovations significantly less attractive. To control for possible between-group confounds, we reran the analysis while controlling for age and education level. As in the previous analysis, the expertise by innovation-type interaction term remained significant at $p < 0.001$, with the same pattern of results. Neither control variable was significant.

Individuals' Characteristics and Perceived Threats

To this point, we have discussed and demonstrated how resistance to innovations may result from the threats that innovations often present to potential adopters. Integrating this point with our main arguments in chapter 1, we propose that perceived threats and individual differences in personality may interact in their effect on resistance to innovation. Specifically, some people may be more sensitive to, and thus more strongly influenced by, perceived threats, in particular threats in the unfamiliar. This possibility is indicated in the moderating effect of individual differences that is drawn in figure 2.2.

As we described in the previous chapter, some people are dispositionally more likely to resist new situations than others. These individuals typically view change in a negative light, feel uncomfortable and even stressed when changes are introduced, and overall have a harder time adjusting to change. It may very well be that for these individuals the perception of threats may have an even more detrimental effect on their reaction to innovations in comparison with individuals with a more favorable disposition toward change. In other words, dispositional resistance to change may moderate the relationship between perceived threats and resistance to innovation such that stronger relationships will emerge for individuals who are dispositionally more resistant to change (see figure 2.8).

To test this relationship, we relied on an earlier study conducted by the first author, using data with a large variety of variables predicting employees' resistance to an organizational restructuring (Oreg 2006). Specifically, one of the variables considered in the Oreg (2006) study was the degree

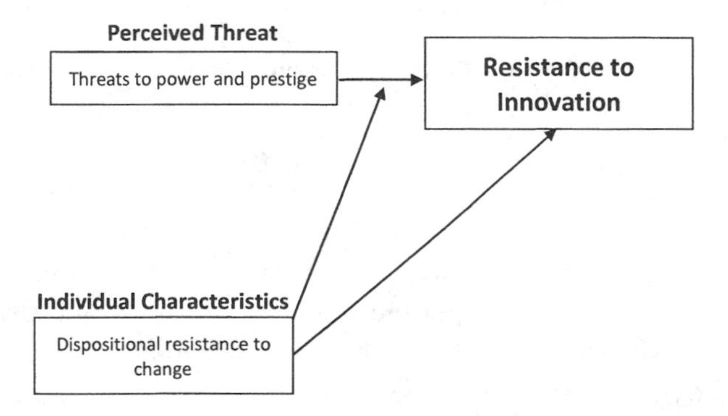

FIGURE 2.8. Expected relationships between dispositional resistance to change, perceived threats, and resistance to innovation.

TABLE 2.2 **Multiple regression analysis with dispositional resistance to change, perceived threats to power and prestige, and their interaction term predicting resistance to change ($N = 175$)**

Variable	B	Std. Err	β
Dispositional resistance to change (RTC)	0.16	0.08	0.14†
Threat to power and prestige	0.39	0.09	0.33**
RTC X Threats to power and prestige	0.17	0.08	0.16*
R^2	0.13		

Note: Variables were centered before including them in the analysis.
†< 0.1.
*$p < 0.05$.
**$p < 0.01$.

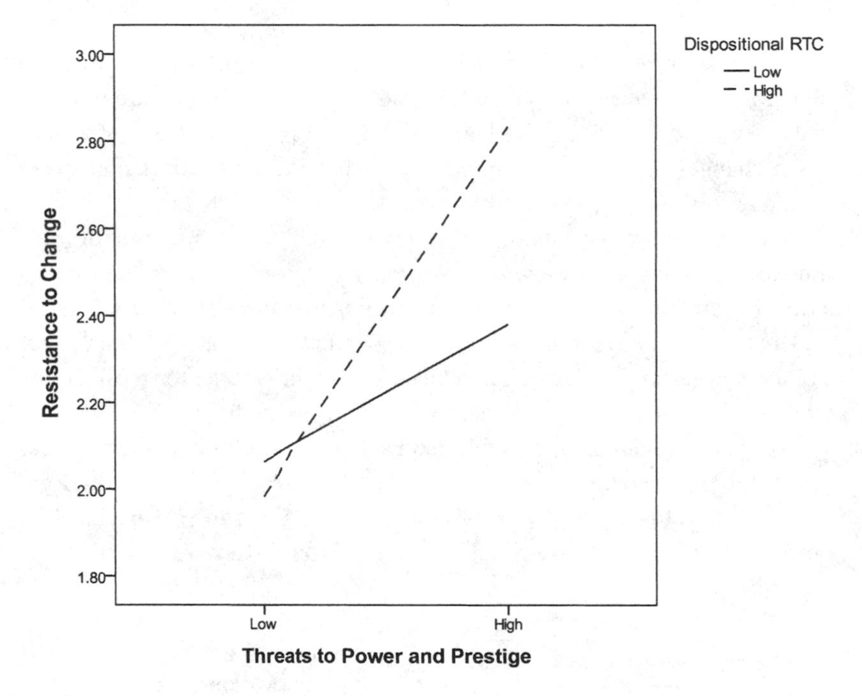

FIGURE 2.9. Moderating effect of dispositional resistance to change on the relationship between perceived threats to power and prestige and resistance to change.

to which employees will perceive the organizational structuring as threatening their political standing in the organization. A second variable involved dispositional resistance to change, assessed with the RTC scale (Oreg 2003), as discussed in the previous chapter. We therefore use these data here for testing the moderation effect proposed above. After centering dispositional resistance to change and perceptions of threats to power and prestige, we included their product in a multiple regression analysis to predict individuals' resistance to the organizational change. Results of this analysis are presented in table 2.2. As can be seen, in support of our moderation hypothesis, the interaction term significantly predicted individuals' resistance to the organizational restructuring. To determine the particular pattern of relationships among variables we plotted the relationship between perceived threats and resistance to change across high and low levels of dispositional resistance to change (see figure 2.9). In line with our expectations, the relationship between perceived threat and resistance to change was stronger among individuals high in dispositional resistance to change.

Conclusions

Our focus in this chapter was on resistance that emanates from the threats that innovations often bring about, with a particular focus on threats to one's gained expertise. Just as innovations have potential benefits, so can they threaten many aspects in individuals' lives. Beyond specific practices that managers and marketers can adopt for dealing with resistance, if one is to accurately interpret consumers' resistance it is important to acknowledge the fact that innovations often incur significant costs for adopters. The most concrete form of threat that many innovations present is to individuals' livelihood, whereby new technology threatens individuals' jobs. For this source of threat, there often isn't much that can be done to alleviate resistance. In the name of efficiency, firms will continue to incorporate new technologies, often with the aim of reducing labor costs. Nevertheless, there are still many cases in which innovations do not serve to substitute, but rather to enhance, individuals' work. Being aware of the possibility that individuals may fear the innovation because of the perceived threat to their job security should behoove marketers and managers to share information about the impending technological change, to assuage undue concerns, and thus reduce resistance. In chapter 3 we discuss in detail this and other factors that have to do with the process of introducing innovations.

Two other related sources of resistance we discussed have to do with individuals' power and expertise. Although distinct, these sources are often related given that power and expertise frequently go hand in hand.

Nevertheless, understanding the difference between the two is important for accurately anticipating and dealing with resistance. Whereas individuals in positions of power fear the loss of status that some innovations may bring about, experts fear losing their competence. From a marketer's perspective this suggests different approaches for circumventing resistance. Where power is concerned, marketers may help potential adopters maintain their position of power by inviting them early on to take part in the development and distribution of the innovation. Such acts of co-optation are often implemented in organizations when preparing for an organizational change (e.g., Gray and Ariss 1985). On the other hand, where threats to expertise and gained skills is the source of threat, means in the form of training workshops and tutorials can be offered for assisting in the transition from previous skills to the new required skills. Our findings with respect to expertise also suggest that in the B2B context, depending on the type of innovation involved (incremental versus radical), marketers may wish to focus their efforts on experts when marketing incremental innovations, yet on novices where radical innovations are concerned.

We concluded our discussion in this chapter by demonstrating the enhancing effect that personality can have on the relationship between perceived threat and reactions to change. Specifically, our findings show that the sense of threat that innovations incur may be particularly detrimental for individuals who are dispositionally predisposed against the notion of change.

In chapter 1 we discussed internal, personality-based sources of resistance, and in this chapter types of threats that innovations may pose. In the following chapter we shift our attention to aspects that may be independent to the innovation itself and, rather, have to do with the manner in which the innovation is presented. In many cases, an innovation that was resisted might have achieved a positive response had it been differently presented.

Notes

1. Although there exists a dispute concerning the degree to which the Dvorak keyboard can actually improve typing speed (Liebowitz and Margolis 1990), the objective performance of the keyboard is irrelevant for our study given that it is the subjective perception of the innovation that will ultimately determine adoption. Regardless of its actual benefits, participants in our study were led to believe that the Dvorak can significantly improve typing speeds.

2. The order of innovations was randomly alternated across respondents. No significant differences were found in responses across the different orderings.

References

Armenakis, Achilles A., Jeremy B. Bernerth, Jennifer P. Pitts, and H. Jack Walker. 2007. "Organizational Change Recipients' Beliefs Scale: Development of an Assessment Instrument." *Journal of Applied Behavioral Science* 43 (4): 495–505.

Ashford, Susan J. 1988. "Individual Strategies for Coping with Stress during Organizational Transitions." *Journal of Applied Behavioral Science* 24 (1): 19–36.

Axtell, Carolyn, Toby Wall, Chris Stride, Kathryn Pepper, Chris Clegg, Peter Gardner, and Richard Bolden. 2002. "Familiarity Breeds Content: The Impact of Exposure to Change on Employee Openness and Well-Being." *Journal of Occupational and Organizational Psychology* 75 (2): 217–31.

Bagozzi, R., F. Davis, and P. Warshaw. 1992. "Development and Test of a Theory of Technological Learning and Usage." *Human Relations* 7 (45): 659–86.

Bedard, Jean, and Michelene T. Chi. 1992. "Expertise." *Current Directions in Psychological Science* 1 (4): 135–39.

Berger, J., and C. Heath. 2008. "Who Drives Divergence? Identity Signaling, Outgroup Dissimilarity, and the Abandonment of Cultural Tastes." *Journal of Personality and Social Psychology* 95 (3): 593.

Bernerth, Jeremy B., Achilles A. Armenakis, Hubert S. Feild, and H. Jack Walker. 2007. "Justice, Cynicism, and Commitment: A Study of Important Organizational Change Variables." *Journal of Applied Behavioral Science* 43 (3): 303–26.

Buhl, Lance C. 1974. "Mariners and Machines: Resistance to Technological Change in the American Navy." *Journal of American History* 61 (3): 703–27.

Caldwell, Steven D., David M. Herold, and Donald B. Fedor. 2004. "Toward an Understanding of the Relationships among Organizational Change, Individual Differences, and Changes in Person-Environment Fit: A Cross-Level Study." *Journal of Applied Psychology* 89 (5): 868–82.

Campanario, J. M. 1996. "Have Referees Rejected Some of the Most-Cited Articles of All Times?" *Journal of the American Society for Information Science* 47 (4): 302–10.

———. 2002. "The Parallelism between Scientists and Students Resistance to New Scientific Ideas." *International Journal of Science Education* 24 (10): 1095–110.

Chase, William G., and Herbert A. Simon. 1973. "Perception in Chess." *Cognitive Psychology* 4 (1): 55–81.

Childers, T. L. 1986. "Assessment of the Psychometric Properties of an Opinion Leadership Scale." *Journal of Marketing Research* 23 (2): 184–88.

Cohen, Wesley M., and Daniel A. Levinthal. 1990. "Absorptive Capacity: A New Perspective on Learning and Innovation." *Administrative Science Quarterly* 35 (1): 128–52.

Coulter, R. A., L. F. Feick, and L. L. Price. 2002. "Changing Faces: Cosmetics Opinion Leadership among Women in the New Hungary." *European Journal of Marketing* 36 (11/12): 1287–308.

David, P. A. 1985. "Clio and the Economics of QWERTY." *American Economic Review* 75 (2): 332–37.

Davis, Fred D., Richard P. Bagozzi, and Paul R. Warshaw. 1989. "User Acceptance of Computer Technology: A Comparison of Two Theoretical Models." *Management Science* 35 (8): 982–1003.

Deci, Edward L., and Richard M. Ryan. 1985. *Intrinsic Motivation and Self-Determination in Human Behavior*. New York: Plenum.

Dent, Eric B., and Susan Galloway Goldberg. 1999. "Challenging 'Resistance to Change.'" *Journal of Applied Behavioral Science* 35 (1): 25–41.

Dewar, Robert D., and Jane E. Dutton. 1986. "The Adoption of Radical and Incremental Innovations: An Empirical Analysis." *Management Science* 32 (11): 1422–33.

Edan, Y. 1995. "Design of an Autonomous Agricultural Robot." *Applied Intelligence* 5 (1): 41–50.

Ericsson, K. Anders, and Jacqui Smith. 1991. "Prospects and Limits of the Empirical Study of Expertise: An Introduction." In *Toward a General Theory of Expertise: Prospects and Limits*, edited by K. Anders Ericsson and Jacqui Smith. New York: Cambridge University Press.

Fedor, Donald B., Steven Caldwell, and David M. Herold. 2006. "The Effects of Organizational Changes on Employee Commitment: A Multilevel Investigation." *Personnel Psychology* 59 (1): 1–29.

Flynn, L. R., R. E. Goldsmith, and J. K. Eastman. 1994. "The King and Summers Opinion Leadership Scale: Revision and Refinement." *Journal of Business Research* 31 (1): 55–64.

———. 1996. "Opinion Leaders and Opinion Seekers: Two New Measurement Scales." *Journal of the Academy of Marketing Science* 24 (2): 137–47.

Fournier, Susan. 1998. "Consumers and Their Brands: Developing Relationship Theory in Consumer Research." *Journal of Consumer Research* 34: 343–73.

———. 2001. "Introducing New Coke." Harvard Business School Case Studies.

———. 2009. "Lessons Learned about Consumers' Relationships with Their Brands." In *Handbook of Brand Relationships*, edited by Deborah J. MacInnis, C. Whan Park, and Joseph R. Priester. Armonk, NY: M. E. Sharpe.

Frensch, Peter A., and Robert J. Sternberg. 1989. "Expertise and Intelligent Thinking: When Is It Worse to Know Better?" In *Advances in the Psychology of Human Intelligence*, edited by R. J. Sternberg. Vol. 5. Hillsdale, NJ: Lawrence Erlbaum Associates.

Fried, Yitzhak, Robert B. Tiegs, Thomas J. Naughton, and Blake E. Ashforth. 1996. "Managers' Reactions to a Corporate Acquisition: A Test of an Integrative Model." *Journal of Organizational Behavior* 17 (5): 401–27.

Gaertner, Karen N. 1989. "Winning and Losing: Understanding Managers' Reactions to Strategic Change." *Human Relations* 42 (6): 527–46.

Goldenberg, Jacob, Oded Lowengart, Shaul Oreg, and Michael Bar-Eli. 2010. "How Do Revolutions Emerge?" *International Studies of Management and Organization* 40 (2): 30–51.

Goldsmith, R. E., and R. Desborde. 1991. "A Validity Study of a Measure of Opinion Leadership." *Journal of Business Research* 22 (1): 11–19.

Goltz, Sonia M., and Amy Hietapelto. 2002. "Using the Operant and Strategic Contingencies Models of Power to Understand Resistance to Change." *Journal of Organizational Behavior Management* 22 (3): 3–22.

Gray, B., and S. S. Ariss. 1985. "Politics and Strategic Change across Organizational Life Cycles." *Academy of Management Review* 10 (4): 707–23.

Hall, D. T., J. G. Goodale, S. Rabinowitz, and M. A. Morgan. 1978. "Effects of Top-Down Departmental and Job Change upon Perceived Employee Behavior and Attitudes: A Natural Field Experiment." *Journal of Applied Psychology* 63 (1): 62–72.

Herold, David M., Donald B. Fedor, and Steven D. Caldwell. 2007. "Beyond Change Management: A Multilevel Investigation of Contextual and Personal Influences on Employees' Commitment to Change." *Journal of Applied Psychology* 92 (4): 942–51.

Hesketh, Beryl. 1997. "Dilemmas in Training for Transfer and Retention." *Applied Psychology: An International Review* 46 (4): 317–39.

Hinds, Pamela J., Michael Patterson, and Jeffrey Pfeffer. 2001. "Bothered by Abstraction: The Effect of Expertise on Knowledge Transfer and Subsequent Novice Performance." *Journal of Applied Psychology* 86 (6): 1232–43.

Hunt, Bruce J. 1983. "'Practice vs. Theory': The British Electrical Debate, 1888–1891." *Isis* 74 (3); 341–55.

Huxley, Thomas Henry, and Leonard Huxley. 1900. *Life and Letters of Thomas Henry Huxley*. London: Macmillan.

Johnson, P. E., A. A. Duran, F. Hassebrock, J. Moller, M. Prietula, P. J. Feltovich, and D. B. Swanson. 1981. "Expertise and Error in Diagnostic Reasoning." *Cognitive Science* 5: 235–83.

Lapointe, Liette, and Suzanne Rivard. 2005. "A Multilevel Model of Resistance to Information Technology Implementation." *MIS Quarterly* 29 (3): 461–91.

Liebowitz, S. J., and Stephen E. Margolis. 1990. "The Fable of the Keys." *Journal of Law and Economics* 33 (1): 1–25.

Lock, S. 1982. "Peer Review Weighed in the Balance." *British Medical Journal* 285 (6350): 1224.

Marchant, Garry, John P. Robinson, Urton Anderson, and Michael Schadewald. 1991. "Analogical Transfer and Expertise in Legal Reasoning." *Organizational Behavior and Human Decision Processes* 48 (2): 272–90.

Maxwell, James Clerk. 1861. "On Physical Lines of Force." *Philosophical Magazine* 21.

———. 1865. "A Dynamical Theory of the Electromagnetic Field." *Philosophical Transactions of the Royal Society of London* 155: 459–512.

———. 1873. *A Treatise on Electricity and Magnetism*. Oxford: Clarendon Press.

McCampbell, Atefeh Sadri. 1997. "Strategy for New Product Development." *Journal of Customer Service in Marketing & Management* 3 (1): 39–58.

Moreau, Page C., Donald R. Lehmann, and Arthur B. Markman. 2001. "Entrenched Knowledge Structures and Consumer Responses to New Products." *Journal of Marketing Research* 37 (8): 14–29.

Morse, N. C., and E. Reimer. 1956. "The Experimental Change of a Major Organizational Variable." *Journal of Abnormal Psychology* 52 (1): 120.

Myers, J. H., and T. S. Robertson. 1972. "Dimensions of Opinion Leadership." *Journal of Marketing Research* 9 (1): 41–46.

Naswall, Katharina, Magnus Sverke, and Johnny Hellgren. 2005. "The Moderating Role of Personality Characteristics on the Relationship between Job Insecurity and Strain." *Work and Stress* 19 (1): 37–49.

Oliver, Thomas. 1986. *The Real Coke, the Real Story*. New York: Penguin Books.

Oreg, Shaul. 2003. "Resistance to Change: Developing an Individual Differences Measure." *Journal of Applied Psychology* 88: 680–93.

———. 2006. "Personality, Context, and Resistance to Organizational Change." *European Journal of Work and Organizational Psychology* 15: 73–101.

Oreg, Shaul, and Jacob Goldenberg. 2007. "Expertise and Resistance to the Adoption of Incremental and Radical Innovations." Paper presented at the 12th International Conference on Quality and Productivity Research.

Patel, Vimla L., David A. Evans, and Guy J. Groen. 1989. "Biomedical Knowledge and Clinical Reasoning." In *Cognitive Science in Medicine: Biomedical Modeling*, edited by D. Evans and V. Patel. Cambridge, MA: MIT Press.

Patel, Vimla L., and Guy J. Groen. 1986. "Knowledge Based Solution Strategies in Medical Reasoning." *Cognitive Science* 10 (1): 91–116.

Paterson, Janice M., and Jane Cary. 2002. "Organizational Justice, Change Anxiety, and Acceptance of Downsizing: Preliminary Tests of an AET-Based Model." *Motivation and Emotion* 26 (1): 83–103.

Paulsen, Neil, Victor J. Callan, Tim A. Grice, David Rooney, Cindy Gallois, Elizabeth Jones, Nerina L. Jimmieson, and Prashant Bordia. 2005. "Job Uncertainty and Personal Control during Downsizing: A Comparison of Survivors and Victims." *Human Relations* 58 (4): 463–96.

Ram, S., and J. N. Sheth. 1989. "Consumer Resistance to Innovations: The Marketing Problem and Its Solutions." *Journal of Consumer Marketing* 6 (2): 5–14.

Reingold, Eyal M., Neil Charness, Richard S. Schultetus, and Dave M. Stampe. 2001. "Perceptual Automaticity in Expert Chess Players: Parallel Encoding of Chess Relations." *Psychonomic Bulletin and Review* 8 (3): 504–10.

Richins, M. L., and T. Root-Shaffer. 1988. "The Role of Involvement and Opinion Leadership in Consumer Word-of-Mouth: An Implicit Model Made Explicit." *Advances in Consumer Research* 15: 32–36.

Rifkin, J. 1995. *The End of Work: The Decline of the Global Labor Force and the Dawn of the Post-Market Era*. New York: G. P. Putnam's Sons.

Rogers, Everett M. 2003. *Diffusion of Innovations*. 5th ed. New York: Free Press.

Rogers, Everett M., and D. G. Cartano. 1962. "Living Research Methods of Measuring Opinion Leadership." *Public Opinion Quarterly* 26 (3): 435–41.

Shapiro, Debra L., and Bradley L. Kirkman. 1999. "Employees' Reaction to the Change to Work Teams: The Influence of "Anticipatory" Injustice." *Journal of Organizational Change Management* 12 (1): 51–67.

Sims, Valerie K., and Richard E. Mayer. 2002. "Domain Specificity of Spatial SExpertise: The Case of Video Game Players." *Applied Cognitive Psychology* 16 (1): 97–115.

Singh, Simon. 2004. *Big Bang: The Origins of the Universe*. New York: Fourth Estate.

Skinner, B. F. 1954. "The Science of Learning and the Art of Teaching." *Harvard Educational Review* 24: 86–97.

Spreitzer, Gretchen M., and Aneil K. Mishra. 2002. "To Stay or to Go: Voluntary Survivor Turnover Following an Organizational Downsizing." *Journal of Organizational Behavior* 23 (6): 707–29.

Summers, J. O. 1970. "The Identity of Women's Clothing Fashion Opinion Leaders." *Journal of Marketing Research* 7 (2): 178–85.

Susskind, A. M., V. D. Miller, and J. D. Johnson. 1998. "Downsizing and Structural Holes: Their Impact on Layoff Survivors' Perceptions of Organizational Chaos and Openness to Change." *Communication Research* 25 (1): 30–65.

Swann, William B., P. J. Rentfrow, and J. S. Guinn. 2003. "Self-Verification: The Search for Coherence." In *Handbook of Self and Identity*, edited by M. Leary and J. Tangney. New York: Guilford Press.

Tichy, Noel M. 1983. *Managing Strategic Change: Technical, Political, and Cultural Dynamics*. New York: Wiley.

Tushman, Michael L., and Philip Anderson. 1986. "Technological Discontinuities and Organizational Environments." *Administrative Science Quarterly* 31 (3): 439–65.

van Dam, Karen. 2005. "Employee Attitudes toward Job Changes: An Application and Extension of Rusbult and Farrell's Investment Model." *Journal of Occupational and Organizational Psychology* 78 (2): 253–72.

Venkatraman, M. P. 1990. "Opinion Leadership, Enduring Involvement and Characteristics of Opinion Leaders: A Moderating or Mediating Relationship." *Advances in Consumer Research* 17 (1): 60–67.

Vieitez, J. C., A. De La Torre Carcía, and M. T. Vega Rodríguez. 2001. "Perception of Job Security in a Process of Technological Change: Its Influence on Psychological Well-Being." *Behaviour and Information Technology* 20 (3): 213–23.

Zaltman, Gerald, and R. Duncan. 1977. *Strategies for Planned Change*. New York: Wiley.

Zauberman, Gal. 2003. "The Intertemporal Dynamics of Consumer Lock-In." *Journal of Consumer Research* 30 (3): 405–19.

CHAPTER THREE

It's Not What You Introduce, It's How You Do It

The Process of Innovation Introduction

The bottle screw cap (see figure 3.1) was patented in 1889 by Dan Rylands. It presented a simple and convenient alternative to cork closures. Although its first documented use was for sealing bottles of whiskey, it gradually made its way as a closure for wine bottles. Following further improvements in its sealing capabilities, the Stelvin screw cap was developed specifically for the wine bottle and has continuously demonstrated its superiority over cork. By the late 1970s several leading Australian wineries released commercial bottling sealed with the Stelvin (Courtney 2001). One wine expert was cited proclaiming:

> The industry loved Stelvin: retailers could stand bottles upright on display shelves, as there was no cork to keep moist. Restaurateurs and event organizers loved Stelvin: a quick flick of the wrist and a bottle was open. Winemakers loved Stelvin because their wines aged slowly and gracefully without the risk of premature oxidisation, which can occur when poor storage conditions allow the cork to dry out. And of course wine makers loved Stelvin as it eliminated the danger of cork taint (Bourne 2000, cited in Mortensen and Marks 2002, 5–6).

Despite all this, however, consumers resisted the Stelvin, and by the early 1980s winemakers reverted to using cork seals. One of the key factors to explain consumers' resistance is their misperception that the Stelvin was an inappropriate closure for quality wines. Despite its current revival, and what seems to be a more successful market penetration, lack of education on consumers' part is still noted as the key source of

FIGURE 3.1. The Stelvin screw cap. Source: AmateurGastronomer.com. Credit: Robin Austin.

resistance to the Stelvin (Barker-Kenny 2010; Stelzer 2003). In a 2002 Australian poll, only 22% of premium wine buyers believed that screw caps preserve wine better than cork (Stelzer 2003). Accordingly, Steve Bell, national sales and export manager at MCG Industries, a market leader in closures for bottles, attributes consumer resistance to a lack of understanding and highlights the role of education as the key to overcoming resistance (Barker-Kenny 2010). It has therefore been recommended that screw cap closure producers take on a more rigorous educational route to persuade winemakers, who in turn should provide more substantial information to their consumers (Barker-Kenny 2010).

It is not yet clear to what extent the screw cap will continue to penetrate the global wine market, including conservative markets such as in France. What is clear is that the resistance it encountered in the 1970s and '80s was not so much because of any disadvantages it may have had, such as the ones discussed in the previous chapter, but rather because of a failure in the *process through which it was introduced*. In other words, resistance resulted from the innovation's presentation process rather than its content. Namely, the screw cap's marketing process (henceforth, process) lacked sufficient communication about the product, which resulted in the failure to combat misperceptions about its quality. Our focus in this chapter will be on a number of process factors that are often responsible for consumer resistance (see figure 3.2). As the highlighted section

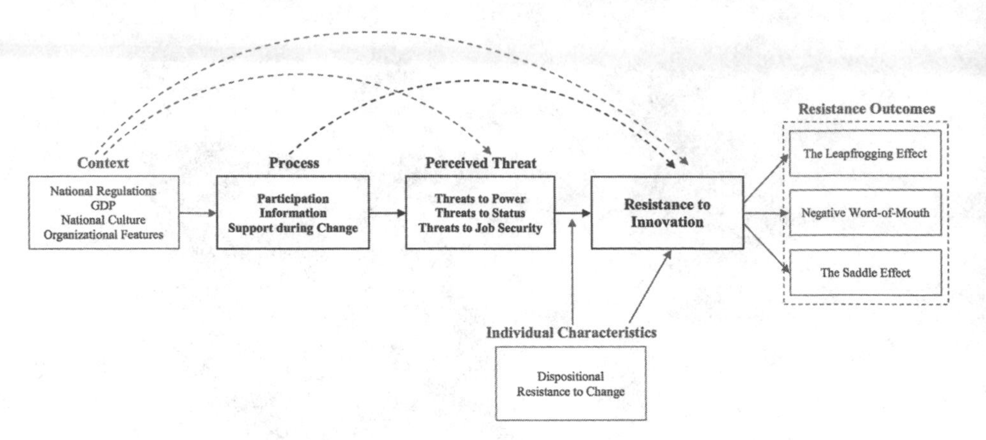

FIGURE 3.2. Change process and resistance to innovation.

in figure 3.2 suggests, beyond demonstrating the direct effect of process on resistance, we also explore the possibility that the effect of process on resistance is mediated by perceived threat.

Indeed, the relative success in reintroducing the Stelvin in Australia in recent years is attributed, to a great degree, to improvement in communicating with customers and educating them about the real nature and advantages of the Stelvin (Garcia, Bardhi, and Friedrich 2007). Helen McGinn, product development manager for Tesco, the United Kingdom's largest wine retailer, talked about the great efforts her company has made in persuading both producers and customers to acknowledge the advantages of the screw cap. She noted that while retailers have noticed an increase in customer willingness to buy expensive wines with screw caps, further educating of the public was still necessary. Accordingly, she said that "when we launch these wines we are going to put out a lot of customer information at the same time. . . . We are really going to make a very big noise about it" (Lechmere 2002, cited in Mortensen and Marks 2002, 11). Some wineries educate both distributors and consumers through information brochures and online videos about the problems with cork seals and advantages of screw caps (Garcia, Bardhi, and Friedrich 2007; see for example videos at http://www.hoguecellars.com/). Another means through which consumer misperceptions about screw caps are being battled is through wine pricing. To turn around consumers' association of screw caps with cheap wines, bottles of wine with screw caps are now often priced higher than the same wine with a cork seal. For example, at

Plumpjack wineries in the Napa Valley, the 1997 Reserve Cabernet was offered for $135 a bottle for the screw cap bottles, whereas bottles sealed with cork were offered at $125 (Garcia, Bardhi, and Friedrich 2007).[1]

The screw cap in the 1970s and '80s is therefore a perfect example of a superior product that was resisted not because of its essence but because of the process through which it was introduced. Systematic research of the role of process in alleviating resistance exists primarily in the context of organizational change. Indeed, the change process is perhaps the most frequently addressed antecedent of resistance in studies of reactions to innovations and change in organizations. In a recent review of empirical studies of reactions to change (Oreg, Vakola, and Armenakis, 2011), forty-two studies involved hypotheses linking the change process to individuals' reactions to the change. Oreg and colleagues classified these studies into categories constituting different aspects of the change process (see table 3.1). Three of these are particularly relevant for the context of consumers' resistance to innovation: (1) lack of communication

TABLE 3.1 **Studies of process variables as predictors of acceptance vs. resistance***

Participation	Amiot et al. 2006; Axtell et al. 2002; Bartunek et al. 1999; Bartunek et al. 2006; Coch and French 1948; Coyle-Shapiro 2002; Daly and Geyer 1994; Eby et al. 2000; Hatcher and Ross 1991; Holt et al. 2007; Korsgaard et al. 2002; Lau and Woodman 1995; Lok et al. 2005; Parsons et al. 1991; Paterson and Cary 2002; Sagie and Koslowsky 1994; Steel and Lloyd 1988; Wanberg and Banas 2000
Communication	Amiot et al. 2006; Axtell et al. 2002; Bordia et al. 2004b; Gaertner 1989; Gopinath and Becker 2000; Johnson et al. 1996; Lau and Woodman 1995; Miller and Monge 1985; Miller et al. 1994; Oreg 2006; Paterson and Cary 2002; Peach et al. 2005; Schweiger and DeNisi 1991; Wanberg and Banas 2000
Management support	Armenakis et al. 2007; Caldwell et al. 2004; Coyle-Shapiro 2002; Eby et al. 2000; Gaertner 1989; Lam and Schaubroeck 2000; Logan and Ganster 2007; Lok et al. 2005; Paterson and Cary 2002; Peach et al. 2005; Wanberg and Banas 2000
Interactional and procedural justice	Armenakis et al. 2007; Armstrong-Stassen 1998; Bernerth et al. 2007; Caldwell et al. 2004; Coyle-Shapiro 2002; Daly and Geyer 1994; Daly 1995; Fedor et al. 2006; Gopinath and Becker 2000; Herold et al. 2007; Korsgaard et al. 2002; Paterson and Cary 2002; Shapiro and Kirkman 1999; Spreitzer and Mishra 2002
Management change competence (in the consumer context this would be equivalent to firm reputation)	Amiot et al. 2006; Rafferty and Griffin 2006

*Adapted from Oreg, Vakola, and Armenakis (2011); full references can be found therein.

and information, (2) lack of participation, and (3) lack of support during the change.

Information and Resistance to Change

Studies on communication and information demonstrate the effects that providing information about change has on individuals' willingness to accept the change (Miller and Monge 1985; Miller, Johnson, Grau 1994; Paterson and Cary 2002; Terry and Jimmieson 2003; Wanberg and Banas 2000). As in the Stelvin case, these studies show that when timely and quality information is provided about the change, individuals become less reluctant and more willing to accept it. Some of the reasons provided for these relationships involve the effect that communication has on change recipients' sense of uncertainty, as well as on their trust in the change agent (e.g., Ashford 1988; Bordia et al. 2004; Schweiger and Denisi 1991).

In one organizational study involving the implementation of new technology, not only information *about* but also actual experience *with* the new technology was tested as a means of decreasing uncertainty and resistance (Axtell et al. 2002). The study took place in a large UK-based distribution company during the introduction of new technology for automatically sorting deliveries by postal area. The new technology was to replace a sequence of old sorting machines that worked at a much slower rate. The new system was also considered much more complex, requiring much more information processing and decision making on the part of the operator. Thus much uncertainty and anxiety were anticipated among employees. Whereas all employees received some initial training with the new system, some had the opportunity for additional experience with it before it actually replaced the old system. Employees provided information about a number of variables, including their openness to the new system and their job-related psychological well-being (anxiety, depression), both before and after its full implementation. As predicted, greater exposure to the new system was associated with improved well-being and greater openness to the new system among the system operators.

Participation and Resistance to Change

A second process variable, receiving perhaps the greatest amount of attention in research on reactions to organizational change, is *participation*.

In many studies, resistance to change was inversely related to the degree to which change recipients had an opportunity to provide their inputs about the change. In a classic organizational study, perhaps the first to empirically consider the construct of resistance to change, Lester Coch and John French (1948) conducted a field experiment in which they tested the impact of employee participation in alleviating resistance to change. Their study was conducted at the Harwood Manufacturing Corporation, in which seamstresses produced pajamas. Every so often, seamstresses were required to adapt to changes in technology or in the structure of their tasks. Following such changes, the company's management observed marked drops in seamstresses' performance and attitude. The researchers, working as consultants at the company, decided to test whether allowing seamstresses to take part in determining the changes would influence their ultimate resistance to the changes.

Their control group went through the company's standard procedures whereby employees were informed of the need to undergo changes in their tasks. Contrarily, experimental groups were given varying degrees of participation in the design of the changes. In some groups representatives provided inputs about the proposed changes and participated in the ultimate job design. In other groups all group members participated in the planning and designing of their new roles. As expected, Coch and French found that as participation increased, employees' acceptance of the change and their performance following it improved. Although some criticism to their methodology has been offered, suggesting that their manipulation of participation was confounded (Bartlem and Locke 1981), to this day Coch and French's study is accepted as a milestone in establishing the link between participation and decreased resistance to change.

Since their seminal study, several other studies confirmed the effects of participation on resistance to change (e.g., Amiot, Jimmieson, and Callan 2006; Coyle-Shapiro 2002; Holt et al. 2007; Steel and Lloyd 1988). Some of these studies indicate that participation lowers resistance because of its positive effect on employees' perception that they can cope with the change (Amiot, Jimmieson, and Callan 2006).

Support during Change and Resistance

A third process variable indicated in Oreg and colleague's (2011) review consists of support during the implementation of change. Given the high levels of uncertainty that exist during the implementation of innovations and change, external support during the transition has been shown to go a long way in alleviating resistances (e.g., Caldwell, Herold, and Fedor 2004; Logan and Ganster 2007; Oreg 2006). For example, in a study of multiple organizations, employees' reports of management support during the change were associated with better adjustment to the change (Caldwell, Herold, and Fedor 2004). Consistent with these findings, many marketers acknowledge the importance of providing customer support in particular following the introduction of new products. This corresponds with the kind of support we proposed in chapter 1 for addressing some consumers' focus on the short-term inconvenience of adopting an innovation. As we noted in chapter 1, when Microsoft's Office 2007 products came out, numerous resources were made available to customers to help them make the transition from the previous Office versions, and similar resources were offered with the transition to Office 2010 and to Office 2013. These resources included video tutorials and elaborate web pages that focused on explaining how old commands can be executed using the new software.

Linking Process to Perceived Utility and Threat

Each of the above process factors can alleviate resistance by decreasing uncertainty about the change. When individuals are given ample information about the change, when they are involved in its planning and implementation, and when the system supports them during the transition, uncertainty about the change is removed, along with the potential resistance. To begin with, uncertainty is linked with resistance because lacking knowledge about a new situation leaves room for individuals to imagine the worst about it. For example, not knowing about the actual attributes of the Stelvin screw cap led many consumers to assume it is inferior to cork, thus inciting their resistance to the Stelvin. In other words, uncertainty about innovations may be associated with perceptions of lower utility and higher threat.

The change process may therefore alleviate resistance by influencing the threat versus utility that individuals perceive in the change. As discussed in the previous chapter, a key factor in explaining resistance to innovation is the perceived threat or utility from the innovation. Innovations that either offer no perceived value, or even worse, that threaten valued factors (such as one's status, power, or expertise), are typically resisted. Such perceived threats may also explain why certain change processes ultimately yield resistance. Perceived utility versus threat may therefore mediate the relationship between the innovation's presentation process and resistance to the innovation (see figure 3.2).

The link between perceived threat and resistance to innovation was established in the previous chapter. There are also a few studies that suggest the link between change/innovation process and perceived threat. For example, in a study of the introduction of new equipment as part of an office automation process, employees were expected to experience threats to their ability to perform their tasks using the new technology (Parsons et al. 1991). For many of them, the new equipment might have meant that they would have to work hard to regain their proficiency in performing tasks. Accordingly, the researchers anticipated that employees would experience varying degrees of skill deficiency. Among the antecedent variables considered in this study was involvement in the planning of the automation process. In line with the rationale presented above, involvement in the automation process was expected to lower the degree of skill deficiency experienced. This was indeed found; employees reporting higher levels of involvement also reported lower levels of perceived skill deficiency.

Similarly, in a study of registered nurses at a large American hospital that implemented a transition to a shared governance management approach, participation in the change process was associated with the perception of greater gains due to the new management approach (Bartunek et al. 2006). By Shared Governance, the authors are referring to "a decentralized approach which gives nurses greater authority and control over their practice and work environment; engenders a sense of responsibility and accountability; and allows active participation in the decision-making process, particularly in administrative areas from which they were excluded previously" (O'May and Buchan 1999, 281, cited in Bartunek et al. 2006, 184).

Thus, there is evidence to support each of the links presented in the highlighted section in figure 3.2. It would be useful for our understanding of the resistance process, however, to demonstrate more explicitly how perceived threat versus utility might mediate between change process and resistance to change. To test this mediation effect, we used two data sets, gathered in the context of two separate projects on the topic of reactions to change. One of these projects involves the study of an organizational restructuring, mentioned in the previous chapter (Oreg 2006).

TABLE 3.2 **Descriptive statistics and correlations among participation, perceived threat to power and prestige, and resistance to change ($N = 177$)**

Variable	Mean	SD	1	2
Participation	1.70	1.04	1	
Threat to power and prestige	2.91	0.74	–0.28**	1
Resistance to change	4.21	1.11	–0.27**	0.55**

$^{**}p < 0.01$; variables were measured on a scale ranging from 1 to 6.

The organization within which the restructuring had taken place was very hierarchical in nature, with a clear chain of command and a high formality of rules. The change consisted of a merger of two of the organization's key units, along with a transformation in one of the units' structure (from a functional to a matrix design). The research model guiding this study included multiple predictors of resistance, including, as described in chapter 2, employees' perceptions that the change will harm their political standing in the organization. Many employees felt that the change will lead to a restriction in the amount of influence they will have in the organization and will hinder their overall status. Indeed, as we described, perceived threat to power and prestige was positively associated with resistance to the change.

Another variable in the data collected, not yet considered in previous publications, had to do with the degree to which employees were involved in the planning and implementation of the change, termed *participation*. We can therefore use these data to test whether perceived threat does in fact mediate the relationship between the change process (in this case, participation) and resistance. Specifically, we expect that perceived threats to employees' power and prestige will mediate the relationship between participation in the change process and resistance to the change. Descriptive statistics for participant demographics and for the three variables in the mediation model are presented in table 3.2.

To test for mediation we followed Reuben Baron and David Kenny's (1986) recommended procedures. These involve (1) demonstrating a relationship between the predictor and mediator, (2) demonstrating a relationship between the mediator and criterion while controlling for the predictor, and (3) comparing the predictor's effect on the criterion before and after including the mediator (Baron and Kenny 1986; MacKinnon, Fairchild, and Fritz 2007). Table 3.2 provides support for the first step, indicating a significant correlation between participation and perceived threat to power and prestige. This finding corresponds with those indicated above about the link between change process and perceived threat. Individuals who participated in the design and implementation of the change perceived it as less threatening to their power and prestige. Table 3.2 also shows that when threat to power and prestige was not controlled for, the correlation between participation and resistance was −0.27 ($p < 0.01$), indicating the expected decreased resistance as participation increased.

Table 3.3 includes results of a multiple regression analysis with participation and threat to power and prestige as predictors of resistance to change, for testing steps 2 and 3 of the mediation analysis. As can be seen, when both the predictor (participation) and mediator (threat to power and prestige) were included in the analysis, the mediator remained significant, whereas the predictor lost significance, indicating full mediation. In other words, these data correspond with the proposition that the alleviating effect of participation on resistance is due to the effect that participation has in removing threat. An effective change process removes the perceived threat, which in turn alleviates resistance to the change.

To test this mediation process in another context, we looked at data about teachers' reactions to a reform implemented in the public education system in Israel. These data were collected as part of a project run by the first author and his colleague, Yair Berson, a part of which appears in Oreg and Berson (2011). The reform was initiated in light of a steady decline in student performance over the past decade. It involved a restructuring of the school system, with implications for factors such as class sizes, criteria for selection of teachers, teachers' work hours and workloads, as well as teachers' salaries and benefits. Although the reform did not include the introduction of any particular innovation, the whole idea of the restructuring, and its details, were quite novel and innovative in

TABLE 3.3 **Multiple regression analysis with participation and threat to power and prestige predicting resistance to change ($N = 177$)**

Variable		B	Std. Err	β
Participation		−0.14	0.07	−0.13
Threat to power and prestige		−0.76	0.10	0.52**
R^2	0.32			

**$p < 0.01$.

the context of the traditional and rather conservative education system in Israel. In response to the introduction of the reform, teachers throughout the country expressed a broad variety of responses, ranging from strong support to fierce resistance.

Our aim in this project was to explain differences in teachers' attitudes toward this change. Among the factors on which we collected data, and which did not pertain to our main interests in the Oreg and Berson (2011) study, were some that are of relevance here. Specifically, with respect to the change process, teachers reported the quality and quantity of information they were given about the change. Information about the perceived threat versus utility about the change was obtained by asking teachers to report the degree to which they believed the change will improve or harm their extrinsic job rewards (such as salary and benefits) and their job prestige. In addition, teachers were asked about their attitude toward the proposed change.

Overall, 586 teachers from seventy-five schools participated in the study, with questionnaires being obtained from five to ten teachers (Mean = 7.81, SD = 1.43) in each school. In line with the dominance of women in the teaching profession in Israel, 72% of respondents were female, 12% were male, and the remaining 16% did not report their sex. Teachers' mean age was 40.50 (SD = 9.18) and their mean tenure (in years) was 15.53 (SD = 9.47).

Given that the project involved a large number of variables, to keep questionnaires at a reasonable length two versions were prepared, with some questions appearing only in one version and other questions only in the other. Specifically, questions about perceived threat versus utility were included in half of the questionnaires, whereas the other half included questions about the quantity and quality of information provided. Thus analyses were conducted at the organization level rather than individual level, by aggregating scores for each of the variables to the school level.[2] These data allowed us to test the mediating effects of two aspects of perceived threat: both threats to extrinsic rewards and threats to prestige.

We followed the same steps as described above for testing mediation. Table 3.4 includes descriptive statistics and correlations among this study's variables. As can be seen, the evaluation of the information provided about the change in each school was negatively associated with the degree to which teachers found the change threatening to both their extrinsic rewards and their prestige. Thus it appears that even though the change may have presented an actual threat to some teachers, the uncertainty

TABLE 3.4 **Descriptive statistics and correlations among information, perceived threat to extrinsic rewards, perceived threat to prestige, and resistance to change (N = 75)**

Variable	Mean	SD	1	2	3
Information	3.72	1.00	1		
Threat to extrinsic rewards	3.26	0.36	−0.28*	1	
Threat to prestige	3.49	0.39	−0.32**	0.52**	1
Resistance to change	3.11	0.45	−0.36	0.41**	0.47**

* $p < 0.05$, ** $p < 0.01$; variables were measured on a scale ranging from 1 to 6.

TABLE 3.5 **Multiple regression analysis with information and threat to extrinsic rewards predicting resistance to change (N = 75)**

Variable		B	Std. Err	β
Information		−0.12	0.05	−0.27*
Threat to extrinsic rewards		0.43	0.14	0.34**
R^2	0.24			

*$p < 0.05$.
**$p < 0.01$.

TABLE 3.6 **Multiple regression analysis with information and threat to prestige predicting resistance to change (N = 75)**

Variable		B	Std. Err	β
Participation		−0.10	0.05	−0.23*
Threat to power and prestige		0.46	0.12	0.40**
R^2	0.27			

* $p < 0.05$.
** $p < 0.01$.

caused by withholding information about the change resulted in a greater perception of threat in comparison with that perceived by those who were given ample information.

Table 3.4 also shows that, as expected, both information and perceived threat were associated with resistance to the change. Schools in which teachers reported receiving more information were less likely to exhibit resistance, and schools in which the perceived threats tended to be higher were more likely to exhibit resistance.

Tables 3.5 and 3.6 display results of multiple regression analyses for testing the mediating role of threats to extrinsic rewards and threats to prestige. As can be seen for both types of threat, when the threat was included in the analysis, the effect of information on resistance decreased.

Contrary to our previous test of mediation (table 3.3), in which the effect of the change process on resistance was no longer significant once controlling for perceived threat (indicating full mediation), in the present case the effect of process (information) remained significant, which suggests partial mediation. In other words, although not the only one, at least one route by which information alleviates resistance is by attenuating the degree to which the change is perceived as threatening.

Conclusions

In this chapter we made the case that the process through which innovations are presented can influence the degree of resistance these innovations will encounter. The findings we described support the notion that an innovation process involving the provision of quality information about the change, opportunities for becoming involved in the planning of the change, and support during the change can help remove much of the uncertainty that typically emerges when innovations are introduced. We further argued that the removal of uncertainty reduces resistance because of its effect in removing adopters' sense of threat from the innovation.

This is not entirely new to marketers. Firms use beta versions, encourage potential adopters to experience the new products, and offer sophisticated support services. These are provided for two main reasons: (1) to collect information and obtain relevant insights that can help improve products, and (2) to accelerate adoption rates.

It is the latter that deals with the issue of resistance. Support is provided not only to alleviate consumers' sense of threat and thus convince them to adopt, but also to learn about the reasons consumers may have for resisting. The goal is to improve firms' understanding of the sources of resistance, and the circumstances in which resistance is exacerbated. Even if resistance cannot be entirely avoided, understanding its sources can uncover alternative routes for approaching resistance-prone consumers. Such learning is important because potential adopters who develop resistance do not necessarily want *more* information. What they need is the right kind of information that accurately relates to the sources of anxiety or specific qualms they may have with respect to the innovation.

By overemphasizing the innovativeness of new products, many firms fail to address other important factors, which can lead to consumer resistance. If we accept the fact that the majority of potential adopters will exhibit at least some form of resistance (see also chapter 7), marketers must be prepared to include a variety of messages in their marketing communications to address the various sources of resistance.

A related observation has to do with the difference between marketing communications for incremental versus radical innovations. Whereas firms invest much in their communications about incremental innovations, they do little to address potential consumer concerns when introducing radical ones. This is despite the fact that radical innovations often elicit even greater resistance than incremental ones. Companies seem to assume that the mere novelty of the radical innovation is sufficient for luring consumers and could therefore substitute for any resistance-alleviating efforts they could have made. Their assumption is that the new product will catch on because of its potential to change people's lives. In some cases this may be true, but in many others people may not necessarily want to have their lives changed.

We should qualify these conclusions, however, and acknowledge that even an informative process may not suffice for removing resistance. In fact, there could be cases in which providing information and opportunities for participation will actually *increase* resistance. This is in the case where threats to adopters are real and meaningful, such that additional information about the innovation will only highlight the potential limitations and restrictions that adopting the innovation will incur. For example, let us return to the case of the distribution company described above (Axtell et al. 2002). The main finding described in that case was that early exposure to the new technology was associated with improved well-being and greater openness to the new system among the system operators. However, whereas exposure to the technology alleviating resistance among prospective users of the technology, Axtell and colleagues also report that it increased the resistance among company managers and engineers. They explain this finding by proposing that the technology may have not lived up to management's expectations, or alternatively, that the new system may have threatened managers' sense of control over their employees given that the new system provided employees with a greater degree of influence and control on the job.

Similarly, in Oreg's (2006) data presented above, whereas participation was associated with lower levels of resistance to the change, additional information about the change was found to *increase* the reported resistance. As in Axtell and colleague's (2002) study, Oreg also explained this finding by suggesting that despite the reduction in uncertainty it offered, the additional information may have also highlighted problems about the change. This finding also demonstrates the distinctiveness of *information* and *participation* and suggests that although they are both aspects of the change process, their effect on recipients' resistance will not always be the same. Whereas participation provides recipients with actual control over the outcomes of the innovation process, information about the innovation does not.

Nevertheless, this is not to suggest that marketers should refrain from involving consumers in the innovation process early on. If resistance is to occur, it is better to emerge at the early stages of the innovation process, when something can still be done to remove possible threats, than after a product is fully designed and launched.

We have thus far covered a number of reasons for which individuals resist innovations. Some have to do with the adopter, others with perceived attributes of the innovation, and its form of presentation. Yet another dimension that may influence an innovation's reception is external to all of these and has to do with the context in which the innovation is presented. This will be our focus in chapter 4.

Notes

1. This is despite the costs of the screw cap being much lower than those of the cork seal. The cork is made out of a natural material harvested from an oak tree known as the Cork Oak (*Quercus suber*). The quality of the cork differs, depending on harvesters' skills, the quality of the oak and its means of processing in the factory. Harvesting cork from oak trees does not harm the trees; however, the trees, which live for 150 years, can only be harvested for cork once every ten years. The first two harvests produce poorer quality oak. The price of the cork is determined by its quality and shape, and ranges from about 10 cents to over 50 cents. In contrast, the cost of the screw cap, which is made in an aluminum factory, averages around 5 cents.

2. Aggregation indexes, such as r_{wg}s (James, Demaree, and Wolf 1993, 306) and ICCs (Bliese 2000), were calculated and supported the appropriateness of aggregating data to the school level.

References

Amiot, C., D. Terry, N. Jimmieson, and V. Callan. 2006. "A Longitudinal Investigation of Coping Processes during a Merger: Implications for Job Satisfaction and Organizational Identification." *Journal of Management* 32: 552–74.

Ashford, Susan J. 1988. "Individual Strategies for Coping with Stress during Organizational Transitions." *Journal of Applied Behavioral Science* 24 (1): 19–36.

Axtell, Carolyn, Toby Wall, Chris Stride, Kathryn Pepper, Chris Clegg, Peter Gardner, and Richard Bolden. 2002. "Familiarity Breeds Content: The Impact of Exposure to Change on Employee Openness and Well-Being." *Journal of Occupational and Organizational Psychology* 75 (2): 217–31.

Barker-Kenny, Berenice. 2010. "The Search for the Perfect Closure Part 2: Screw Caps," *Vintage Direct*.

Baron, Reuben M., and David A. Kenny. 1986. "The Moderator-Mediator Variable Distinction in Social Psychological Research: Conceptual, Strategic, and Statistical Considerations." *Journal of Personality and Social Psychology* 51 (6): 1173–82.

Bartlem, Carleton S., and Edwin A. Locke. 1981. "The Coch and French Study: A Critique and Reinterpretation." *Human Relations* 34 (7): 555–66.

Bartunek, Jean M., Denise M. Rousseau, Jenny W. Rudolph, and Judith A. DePalma. 2006. "On the Receiving End: Sensemaking, Emotion, and Assessments of an Organizational Change Initiated by Others." *Journal of Applied Behavioral Science* 42 (2): 182–206.

Bliese, Paul D. 2000. "Within-Group Agreement, Non-Independence, and Reliability: Implications for Data Aggregation and Analysis." In *Multilevel Theory, Research, and Methods in Organizations: Foundations, Extensions, and New Directions*, edited by Katherine J. Klein and Steve W. J. Kozlowski. San Francisco: Jossey-Bass.

Bordia, Prashant, Elizabeth Hunt, Neil Paulsen, Dennis Tourish, and Nicholas DiFonzo. 2004. "Uncertainty during Organizational Change: Is It All about Control?" *European Journal of Work and Organizational Psychology* 13 (3): 345–65.

Caldwell, Steven D., David M. Herold, and Donald B. Fedor. 2004. "Toward an Understanding of the Relationships among Organizational Change, Individual Differences, and Changes in Person-Environment Fit: A Cross-Level Study." *Journal of Applied Psychology* 89 (5): 868–82.

Coch, Lester, and John R. P. French Jr. 1948. "Overcoming Resistance to Change." *Human Relations* 1: 512–32.

Courtney, S. 2001. "The History and Revival of Screw Caps." *Wine of the Week*.

Coyle-Shapiro, Jacqueline A. M. 2002. "Changing Employee Attitudes: The Independent Effects of TQM and Profit Sharing on Continuous Improvement Orientation." *Journal of Applied Behavioral Science* 38 (1): 57–77.

Garcia, R., F. Bardhi, and C. Friedrich. 2007. "Overcoming Consumer Resistance to Innovation." *MIT Sloan Management Review* 48 (4): 82.

Holt, Daniel T., Achilles A. Armenakis, Hubert S. Feild, and Stanley G. Harris. 2007. "Readiness for Organizational Change: The Systematic Development of a Scale." *Journal of Applied Behavioral Science* 43 (2): 232–55.

James, Lawrence R., Robert G. Demaree, and Gerrit Wolf. 1993. "R_{wg}: An Assessment of Within-Group Interrater Agreement." *Journal of Applied Psychology* 78 (2): 306.

Logan, Mary S., and Daniel C. Ganster. 2007. "The Effects of Empowerment on Attitudes and Performance: The Role of Social Support and Empowerment Beliefs." *Journal of Management Studies* 44 (8): 1523–50.

MacKinnon, David P., Amanda J. Fairchild, and Matthew S. Fritz. 2007. "Mediation Analysis." *Annual Review of Psychology* 58: 593–614.

Miller, Katherine I., and Peter R. Monge. 1985. "Social Information and Employee Anxiety about Organizational Change." *Human Communication Research* 11 (3): 365–86.

Miller, V. D., J. R. Johnson, and J. Grau. 1994. "Antecedents to Willingness to Participate in a Planned Organizational Change." *Journal of Applied Communication Research* 22 (1): 59–80.

Mortensen, W., and B. Marks. 2002. "An Innovation in the Wine Closure Industry: Screw Caps Threaten the Dominance of Cork." Victoria University of Technology Working Paper Series 18, 1–14.

O'May, F., and J. Buchan. 1999. "Shared Governance: A Literature Review." *International Journal of Nursing Studies* 36 (4): 281–300.

Oreg, Shaul. 2006. "Personality, Context, and Resistance to Organizational Change." *European Journal of Work and Organizational Psychology* 15: 73–101.

Oreg, Shaul, and Yair Berson. 2011. "Leadership and Employees' Reactions to Change: The Role of Leaders' Personal Attributes and Transformational Leadership Style." *Personnel Psychology* 64 (3): 627–59.

Oreg, Shaul, Maria Vakola, and Achilles A. Armenakis. 2011. "Change Recipients' Reactions to Organizational Change: A Sixty-Year Review of Quantitative Studies." *Journal of Applied Behavioral Science* 47 (4): 461–524.

Parsons, Charles K., Robert C. Liden, Edward J. O'Connor, and Dennis H. Nagao. 1991. "Employee Responses to Technologically Driven Change: The Implementation of Office Automation in a Service Organization." *Human Relations* 44 (12): 1331–56.

Paterson, Janice M., and Jane Cary. 2002. "Organizational Justice, Change Anxiety, and Acceptance of Downsizing: Preliminary Tests of an AET-Based Model." *Motivation and Emotion* 26 (1): 83–103.

CHAPTER FOUR

Where and When Is the Innovation Introduced?

The Role of Innovation Context in the Emergence of Resistance

"Our coffeehouses have become a beacon for coffee lovers everywhere. Why do they insist on Starbucks? Because they know they can count on genuine service, an inviting atmosphere and a superb cup of expertly roasted and richly brewed coffee every time." This blurb appears on Starbucks' website and describes what is probably considered true in many parts of the globe. Starbucks has penetrated the coffee market in more than fifty countries, with more than eleven thousand locations in the United States alone. The company attribute its success to its mission of "inspiring and nurturing the human spirit." The key to its success has been said to rely on the innovative notion of creating a "third place between work and home," by combining emphases on an excellent product, a pleasing physical environment, and service-minded employees (Berry et al. 2006). The variety of coffee drinks offered was particularly innovative in comparison to what was offered in many small town cafés in the United States, in which the coffee experience consisted of filter coffee on a hot plate (Patterson, Scott, and Uncles 2010).

Nevertheless, Starbucks' innovative idea has not achieved success everywhere. Some examples of Starbucks' failures include Australia, in which the majority of Starbucks shops were shut down in 2008, and Israel from which Starbucks retreated entirely in 2003. Paul Patterson, Jane Scott, and Mark Uncles (2010) explain "how the local competition [in Australia] defeated a global brand." Among the explanations they provide, they suggest that whereas Starbucks coffee presented a level of sophistication with which most Americans were not familiar, the Australian coffee market has been mature and sophisticated for decades. Furthermore, whereas Starbucks assumes that people's decision to visit will be determined by convenience in location, Australians are willing to travel out of their way for their favorite cup of coffee. As Patterson and colleagues put it, "For Australians, coffee is as much about relationships as it is about the product, suggesting that an impersonal, global chain experience would have trouble replicating the intimacy, personalization, and familiarity of a suburban boutique café" (Patterson, Scott, and Uncles 2010, 43). Moreover, given Australians' already sophisticated palates, what is considered in America to be great coffee was seen in Australia as a "'gimmicky,' 'watered down product,' consisting of 'buckets of milk'" (44). Such factors, along with the already rough competition in the Australian coffee market, led Starbucks to close sixty-one of their eighty-four stores in July 2008. In sum, what Patterson and colleagues suggest is that Starbucks simply failed to adapt its product and services to the local culture. It overestimated the value that the Australian consumer will attach to the in-store experience, and underestimated the relational aspect of the coffee experience in Australia and the different definition that quality coffee has in this new market.

A similar process explains why Starbucks failed in Israel. As one reporter put it in the headline of his article, "For America, the coffee virgin, Starbucks is the first man and the perfect lover. For us [in Israel], it is no more than a pimply boy, trying to interest us with his dubious graces" (Shaked 2003). Indeed, it took Starbucks only two years from opening their six shops in Israel in 2001 to closing all six in April 2003. The Israeli Starbucks managers acknowledged that they simply failed to adapt to the local coffee culture (Coussin 2003). As in Australia, Israeli coffee culture, with strong European influences, was highly developed well before Starbucks showed up. Leading Italian brand names such as Segafredo, Lavazza, and Illy had been familiar to the average Israeli coffee consumer for years, and numerous local coffee chains had been flourishing in Israel, and Tel-Aviv in particular, by the mid-1990s. The relative advantages that Starbucks had in many of the countries it entered, of superior technology, high-quality products, and high-quality service, were not considered advantages in Israel. The local coffee technology was not inferior to that of Starbucks, high-quality service was already prevalent in the Israeli coffee market, and Starbucks coffee didn't appeal to the Israeli tastes.

Patterson and colleagues (2010) note the irony in the fact that whereas Starbucks' success originated in its ability to adapt the European coffee

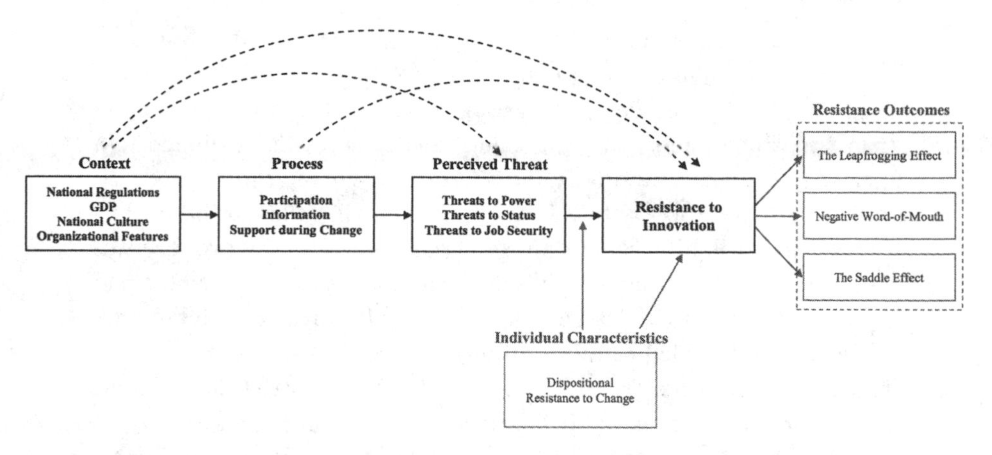

FIGURE 4.1. Context and resistance to innovation.

experience to the American market, it failed to adapt their US product and experience to several other local cultures. What Starbucks did (the change content), and how it did it (the change process) were virtually the same across the globe. It is the *context* that differed each time, and in some cases, as in those of Australia and Israel, made all the difference. As we will elaborate in this chapter, even when an innovation's *content* (its perceived benefit versus threat) and *process* appear to be conducive for adoption, even for seemingly similar markets, innovations may still be resisted because of the particular *context* in which they are introduced (see figure 4.1).

Failures of Particular Innovations in Particular Contexts

In *context* we are referring to anything in the environment or setting within which an innovation is introduced. These include the geographical, political, and social settings, which vary across locations and over time. A convenient way to examine the effect of context is by looking at cross-national differences in the innovation adoption process (although context will obviously vary even within nations). Indeed, numerous cross-national studies have been conducted involving a large variety of contextual variables. We will review these briefly and devote particular attention to the effects of culture on employees' resistance to innovation.

As in the Starbucks case, there are numerous others in which an innovation that did extremely well in some countries failed in others. One such other case is that of Wal-Mart's entry into Germany (Zinkhan et al.

2009). Founded in 1962, Wal-Mart has become the largest private employer and the largest retailer in the United States (Basker 2007) as well as the largest food retailer in the world (Zinkhan et al. 2009), with more than 8,700 retail units in fifteen countries. Its success is attributed to a great degree to the innovative and pioneering design of its logistics, distribution, and inventory control. It was among the first chain to have all its stores and distribution centers hooked up to a computer network. It was also among the first to adopt bar-code technology, which allowed them to cut labor costs by 50% (Basker 2007). In 1990 it introduced Retail Link software that provided detailed inventory data to all stores, distribution centers, and suppliers. The use of innovative technology to facilitate coordination among its units and reduce labor costs continues to be one of its primary edges over its competitors. Wal-Mart operates the world's largest private satellite communications network, which allows it to even regulate the temperature in individual stores and has considered employing Radio Frequency Identification technology, which makes tracking shipments, inventory, and sales much easier by assigning each item with a unique tag that can be read by radio signal (Basker 2007). Beyond these technological advantages, its ability to use economies of scale to continuously lower prices further amplifies its advantages over its competitors.

These advantages, however, were not realized in all of the markets Wal-Mart entered. In 2006, after eight years in which it opened eighty-five stores in Germany, Wal-Mart announced that it would be closing its businesses in the country, incurring a loss of $1 billion. The main reason provided by analysts was Wal-Mart's failure to adapt to the local environment. Even before Wal-Mart's decision to close, in 2003, Knorr and Arndt wrote an article titled "Why Did Wal-Mart Fail in Germany?" In its abstract they write: "[Wal-Mart] pursued a fundamentally flawed internationalization strategy due to an incredible degree of ignorance of the specific features of the extremely competitive German retail market . . . instead of attracting consumers with an innovative approach to retailing, as it has done in the USA, in Germany the company does not seem to be able to offer customers any compelling value proposition in comparison with its local competitors" (Knorr and Arndt 2003).

Chief in the explanations provided for Wal-Mart's failure were the arrogance of its management and their ignorance of many aspects of Germany's culture and standards, their disregard of German laws and regulations, and their failure to deliver on Wal-Mart's legendary commitments to "sell for less—always" and to provide superior customer service.

First, it took Wal-Mart more than two years before they decided to appoint a German CEO to head the company in Germany. Although the official corporate language in Wal-Mart Germany was English, and both the CEO and most of the company executives in Germany spoke only English, many of the older managers Wal-Mart hired in Germany did not speak English. Consequently, problems in communication often resulted in employee turnover, as well as the loss of key business connections. Several key suppliers refused to work with Wal-Mart, which subsequently led to a restriction in the range of goods the company could offer (Zinkhan et al. 2009). Furthermore, whereas much of Wal-Mart's savings in the United States come from the effectiveness of its innovative technologies in streamlining the logistics chain between producer and retailer, this competitive advantage was substantially restricted in Germany because of the already short logistics chain, compared with the chain in the United States (Christopherson 2007).

Second, company executives were either ignorant of or apathetic about the legal and institutional framework of the German retail market. Contrary to the nonunionized setting of Wal-Mart in the United States, a large portion of the German retail workforce is still unionized, and unions are still extremely influential. Following the need to cut costs early on, Wal-Mart let go of many of its employees, blatantly ignoring Germany's strict worker protection regulations. Much to Wal-Mart's surprise, these actions resulted in union-organized walkouts at thirty stores throughout the country, which resulted in both lost sales and bad publicity (Christopherson 2007; Knorr and Arndt 2003). Furthermore, several German laws directly negated many aspects of Wal-Mart's standard strategy for obtaining market superiority. For example, shopping hour regulations in Germany prohibit stores from operating on Sunday and holidays and allow for a maximum of eighty hours of operation per week. This prevented Wal-Mart from providing the convenience they typically offer outside of Germany, which contributes to customers' positive service experience. Other German regulations, involving antitrust legislation, prohibited Wal-Mart from offering products below cost prices—a tactic the company has often employed for obtaining a competitive advantage. As with respect to Germany's labor laws, here too Wal-Mart repeatedly disregarded German law, which incurred both hefty fines and bad publicity for the company (Knorr and Arndt 2003).

Third, whereas Wal-Mart's use of economies of scale allows it to live up to its promise of providing the best-available value almost everywhere else, the situation in Germany was such that local competitors, such as the retailers Aldi and Rewe, have already occupied the "hard discounter" niche, "typically offering a range of 600 to 700 products, with a high share of own-brands, at rock bottom price and ultra-low margins" (Knorr and Arndt 2003, 15). In addition, whereas one means of cutting costs at Wal-Mart has been by bypassing wholesaler intermediaries and dealing directly with factories, wholesalers in Germany managed to maintain their role as intermediaries between food producers and food distributors, forcing the company to take on the extra costs it typically manages to avoid in the United States (Christopherson 2007). In this context Wal-Mart repeatedly failed to offer lower prices.

Furthermore, the company failed to live up to its promise of providing "excellent customer service." For example, among Wal-Mart's fixed employee guidelines was the requirement that clerks smile and offer to bag customers' groceries. While successfully providing a satisfying customer experience elsewhere, these practices were perceived in Germany as annoying and distasteful (Landler 2006).

Thus innovative strategies that have proven successful for Wal-Mart in many contexts failed miserably in Germany. Several company executives acknowledged that mistakes were made in adapting the company's culture and practices to the local environment in Germany. Knorr and Arendt (2003) more bluntly suggested that Wal-Mart's entry into Germany presents a "textbook case [of] how *not* to enter a foreign market" (5, emphasis in original).

Many other cases depict a similar pattern as that exemplified in both the Starbucks and Wal-Mart cases. Whirlpool's World Washer, designed for emerging markets, was well received in practically all of the countries in which it was launched, with the exception of India (Chavan et al. 2009). When the company looked into the matter, it learned that the problem involved a millimeter-wide gap between the machine's agitator and drum that went unnoticed all over the world, yet created serious problems when Indians tried to wash their paper-thin saris. These got caught in the thin gap in the machine, and by the end of the wash were shredded. Given the company's need to restructure its business model and redesign a new machine for India, it took years for it to recoup its losses and regain a meaningful market share in India.

Another marketing blunder in India is that of Kellogg's Corn Flakes. These were resisted in India because they weren't compatible with Indians' breakfast habits (Chavan et al. 2009; Haig 2005). The locals in India are used to eating warm foods in the morning. They like their milk warm, and sweet. So when they poured boiling milk into their bowl of corn flakes, the flakes became soggy and unappealing. When they tried them with cold milk, the sugar they added wouldn't dissolve, and the flakes weren't sweet enough. Thus one of the most common breakfasts in America failed in a country where cultural eating habits were not compatible with the product. Conversely, Rivella, also known as "the quintessentially Swiss soft drink" (Mirza 2010) failed in the United States. Rivella is a milk-derived drink, produced in Switzerland since 1952. Whereas the drink is extremely popular in Switzerland, to the degree that Rivella spokesperson Monika Christener suggested that the Swiss are "are almost as familiar with it as breast milk" (cited in Mirza 2010), it was pulled from the US market after less than six months because Americans were put off by the idea of a milk-derived soft drink.

All of these cases demonstrate the notion that the benefit of an innovation cannot be judged independently of the context within which it is introduced. Innovations that are embraced in a given setting may very well be incompatible with aspects of another setting, and are consequently resisted. The issue in all of these cases is a matter of fit. Fit between the features of the innovation with features of the context within which it is introduced. Alongside this role of context, one may also consider the possibility that some contexts are less receptive to innovations overall, across types of innovation. The introduction of innovations in these contexts is typically less likely to succeed than in other contexts, independent of the particular nature of the given innovation. In fact, this role of context has been most rigorously addressed in the academic literature. In the following section, we discuss a variety of contextual variables that have been linked with the overall inclination to resist innovations. Although in some of these studies the argument made is that context influences the adoption of particular kinds of products, in most studies it is proposed that the impact of context is expected across innovation types.

Cross-National Differences in Innovation Adoption

Numerous studies compared adoption rates of innovations across countries. As discussed in a number of reviews (e.g., Dekimpe, Parker, and Sarvary 2000; Peres, Muller, and Mahajan 2010), these studies considered a variety of country-level variables, including economic status, population density, culture, and several other societal factors. Findings from a subset of these studies are summarized in table 4.1.

TABLE 4.1 **Selected studies of country-level variables as predictors of innovation adoption/resistance**

Context antecedent category	Specific variable/ dimension	Studies	No. of countries in study	Finding
Timing of innovation launch	*Lead-lag*	(Kumar et al. 1998)	14	In countries that were "lagged" in the introduction of the product, diffusion of consumer electronic products, such as microwaves, VCRs, and personal computers, was faster.
		(Takada and Jain 1991)	5	Diffusion rate of consumer durable goods in Pacific Rim countries was faster in countries with a "lagged introduction" (countries in which the product was introduces after it had already been introduced in other countries).
Societal variables	*Cosmopolitanism*	(Gatignon et al. 1989)	14	Cosmopolitanism had a positive effect on the population's propensity to innovate (rate and level of diffusion of consumer durables, such as freezers, lawn mowers and TVs).
		(Kumar et al. 1998)	14	Cosmopolitanism had a positive effect on innovation diffusion (product sales) of consumer electronics (e.g., VCRs, cell phones).
	Women's participation in the labor force	(Gatignon et al. 1989)	14	Percentage of women participating in the labor force was related to the diffusion of innovations. The effect differed, however, by type of innovation, with a positive effect for products where women play a central communication role in their diffusion, such as washing machines, and a negative effect for products where women do not play such a communication role, such as color televisions and lawnmowers.
		(Kumar et al. 1998)	14	Women's participation in the labor force had a negative effect on diffusion (product sales) of consumer electronics.
	Population density	(Haapaniemi and Mäkinen 2008)	49	Population density and illiteracy rate moderated the effect of culture on innovation takeoff time of PCs and internet hosts.
		(Stump et al. 2010)	76	Population density had a negative effect on the adoption and use of mobile phones.
	Religiosity	(Chandrasekaran and Tellis 2008)	40	Takeoff of new products (e.g., washing machines, MP3 players) was slower among countries with higher religiosity rates.
Economic variables	*Wealth*	(Chandrasekaran and Tellis 2008)	40	Wealth had a positive effect, and disparity (as measured with the Gini index) a negative effect, on takeoff time of new products (consumer electronics).

		(Haapaniemi and Mäkinen 2008)	49	Wealth (GDP) moderated the effect of culture on takeoff time of technology innovation. Only in wealthy countries (with high GDP) cultural characteristics, in particular uncertainty avoidance, was related to innovation adoption. This is because in countries with low GDP, in which people can't afford to buy certain products (e.g., cellular phones, PCs and Internet hosts), there is little variance to begin with that could be predicted by cultural dimensions.
		(Park 2001)	45	National income had a positive effect on Internet use.
		(Stremersch and Lemmens 2009)	34	Economic wealth had a positive effect on sales of new drugs.
		(Stremersch and Tellis 2004)	15	Wealth (GDP) had a positive effect on sales growth of consumer durables (e.g., DVDs, CD players, refrigerators).
		(Stump et al. 2010)	75	Wealth (GDP) had a positive effect on the adoption of mobile phones.
		(Tellis et al. 2003)	15	Wealthier countries exhibited faster takeoff for consumer durables.
	Developed vs. developing countries	(Chandrasekaran and Tellis 2008)	40	Takeoff for products was faster in developed, in comparison to developing, countries. In newly developed Asian countries takeoff was faster than that in industrialized European countries.
	Life style	(Helsen et al. 1993)	12	Lifestyle (standard of living) and health status had a positive and systematic effect on the diffusion of consumer durables.
Hofstede's cultural dimensions	*Uncertainty avoidance*	(Chandrasekaran and Tellis 2008)	40	Faster takeoff time for consumer electronics was exhibited in countries high on uncertainty avoidance. This effect, however, was only established in bivariate analyses. Multivariate analyses, with all of the study variables included, yielded the opposite effect for uncertainty avoidance.
		(Haapaniemi and Mäkinen 2008)	49	Countries with high uncertainty avoidance exhibited longer takeoff for mobile phones.
		(Nilsson 2007)	2	Uncertainty avoidance was said to be negatively associated with use of self-service technologies. Specifically, in Estonia, which is high on uncertainty avoidance, use of internet banking services was exhibited primarily by a restricted sector of society, whereas they were used by a heterogeneous sample, with a broader range of demographics, in Sweden, which is lower on uncertainty avoidance.
		(Park 2001)	45	Uncertainty avoidance had a negative effect on internet use among individuals from 45 countries.*

(*continues*)

TABLE 4.1 ***(continued)***

Context antecedent category	Specific variable/ dimension	Studies	No. of countries in study	Finding
		(Png et al. 2002)	24	Uncertainty avoidance was negatively associated with companies' likelihood of adopting new IT infrastructure.
		(Steenkamp et al. 1999)	11	Uncertainty avoidance was negatively associated with consumer innovativeness, measured with the consumer-specific exploratory acquisition of products scale.
		(Stump et al. 2010)	76	Uncertainty avoidance had a negative effect rate of adoption and use of mobile phones.
		(Tellis et al. 2003)	16	Uncertainty avoidance was negatively related to takeoff of consumer durables.
		(Van Everdingen and Waarts 2003)	10	Uncertainty avoidance had a negative effect on ERP (IT-based innovation) penetration (adoption rate) in medium-sized companies
	Power distance	(Chandrasekaran and Tellis 2008)	40	Takeoff of new consumer electronics took longer in countries higher on power distance.
		(Dwyer et al. 2005)	13	Power distance was positively associated (marginally significant) with the diffusion of technological innovations (durables) across Europe.
		(Haapaniemi and Mäkinen 2008)	49	Takeoff for mobile phones and internet hosts took longer in countries with higher levels of power distance.
		(Singh 2006)	2	Consumer innovativeness (operationalized as consumer-specific exploratory acquisition of products) was higher in Germany, which is lower on power distance, than France, which scores higher on this dimension.
		(Van Everdingen and Waarts 2003)	10	Power distance had a negative effect on ERP (IT-based innovation) penetration (adoption rate) in medium-sized companies.
	Individualism-Collectivism	(Chandrasekaran and Tellis 2008)	40	Countries characterized with in-group collectivism had a slower takeoff time of new products than countries low on in-group collectivism (corresponding with a positive effect of individualism).
		(Dwyer et al. 2005)	13	Individualism was negatively associated with the diffusion rate of technological innovations.
		(Haapaniemi and Mäkinen 2008)	49	Takeoff time of technology (cellular phones, PCs and Internet hosts) was shorter in individualist countries.

	(Park 2001)	45	Individualism had a positive effect on Internet use among individuals from 45 countries.
	(Steenkamp et al. 1999)	11	Individualism was positively associated with consumer innovativeness, measured with the consumer-specific exploratory acquisition of products scale.
	(Stremersch and Lemmens 2009)	34	Individualism was positively associated with country-level sales of new drugs.
	(Stump et al. 2010)	76	Individualism had a negative effect on the growth rate of mobile phones.
	(Van Everdingen and Waarts 2003)	10	Individualism had a positive effect on ERP (IT based innovation) penetration (adoption rate) to medium-sized companies (individualist companies tended to adopt ERP faster).
Masculinity-Femininity	(Dwyer et al. 2005)	13	Masculinity was positively associated with diffusion rate of technological innovations.
	(Park 2001)	45	Masculinity had a negative effect on use of Internet services.
	(Singh 2006)	2	Consumer innovativeness (operationalized as consumer-specific exploratory acquisition of products) was higher in Germany, in which scores on masculinity are relatively high, than France, which scores lower on masculinity.
	(Steenkamp et al. 1999)	11	Masculinity was positively associated with consumer innovativeness, measured with the consumer-specific exploratory acquisition of products scale.
	(Stremersch and Lemmens 2009)	34	Masculinity was negatively associated with country-level sales of new drugs.
	(Van Everdingen and Waarts 2003)	10	Masculinity had a negative effect on the penetration of ERP systems to medium-sized companies.
Long-term orientation	(Dwyer et al. 2005)	13	Long-term orientation was negatively associated with the diffusion rate of technological innovations in Europe.
	(Stremersch and Lemmens 2009)	34	Long-term orientation had a negative effect on sales of new drugs.
	(Van Everdingen and Waarts 2003)	10	Long-term orientation had a positive effect on the penetration of ERP (IT based innovation) penetration (adoption rate) to medium-sized companies

A common contextual variable to have been considered is that of products' timing of introduction with respect to their introduction in other countries, termed *lead-lag* (see Peres, Muller, and Mahajan 2010 for a review of these studies). As indicated in the sample studies included in table 4.1, diffusion has been shown to be faster when innovation introduction is lagged, thus indicating that resistance to innovation will be higher in contexts that are the first to host the innovation. In other studies, national wealth and prosperity, as indicated by variables such as GDP, population density, and average standard of living, were positively associated with innovation adoption, indicating greater resistance in less prosperous countries (e.g., Chandrasekaran and Tellis 2008; Haapaniemi and Mäkinen 2008; Helsen, Jedidi, and DeSarbo 1993).

Other societal factors to have been linked with innovation adoption include cosmopolitanism (Gatignon, Eliashberg, and Robertson 1989; Kumar, Ganesh, and Echambadi 1998), religiosity (Chandrasekaran and Tellis 2008), and sex roles, as indicated in the percentage of women's participation in the work force (Gatignon, Eliashberg, and Robertson 1989; Kumar, Ganesh, and Echambadi 1998). With respect to cosmopolitanism and religiosity, findings have been consistent in showing that national conservatism (low cosmopolitanism and high religiosity) is associated with greater resistance to innovation (slower adoption rates). Findings about the effects of women's participation in the labor force, however, have been inconsistent, with the effect of women's participation depending on the type of innovation involved. At least in one study, the effect of women's participation in the labor force was positive for products with respect to which women play a central communication role in their diffusion, such as washing machines, and negative for products with respect to which women do not play such a role, such as color televisions and lawn mowers (Gatignon, Eliashberg, and Robertson 1989).

Culture is among the most frequently used country-level variable to have been studied to explain cross-country differences in innovation adoption. A number of cultural taxonomies have been used, with a large portion of diffusion studies employing Geert Hofstede's cultural dimensions (Hofstede 2001; 1980). Through a study of employees from more than seventy countries, Hofstede (2001) proposed five dimensions on which countries could be compared: uncertainty avoidance, power distance, individualism (versus collectivism), masculinity (versus femininity), and long-term (versus short-term) orientation. Since their introduction, these cultural dimensions have been linked with numerous country-level outcomes, including several related to the adoption of innovations.

In several studies, *uncertainty avoidance* was linked with greater resistance to innovation, as observed in slower innovation diffusion rates (see table 4.1). According to Hofstede, uncertainty avoidance explains how societies deal with unknown aspects of the future. Individuals in societies that are low on uncertainty avoidance are more inclined to take risks and are tolerant of novel ideas. In contrast, in societies high on uncertainty avoidance, individuals are more anxious about the future. They tend to be intolerant toward ambiguity, stick to old and familiar patterns of behavior, and are therefore more likely to exhibit resistance toward innovations. As can be seen in table 4.1, this tends to be the pattern found in several cross-cultural studies. For example, using data from seventy-six countries, the adoption growth rates of mobile phones across a five-year period correlated negatively with countries' uncertainty avoidance (Stump, Gong, and Chelariu 2010). Similarly, in a multinational survey of companies from twenty-four countries, businesses from countries higher on uncertainty avoidance were less likely to adopt new IT infrastructure (Png, Tan, and Wee 2002).

Interestingly, however, a number of researchers suggested that the relationship between uncertainty avoidance and innovation adoption should be *positive* rather than negative, in particular for technological advances (e.g., Chandrasekaran and Tellis 2008). The rationale of this explanation is that technology is pursued as a means of avoiding uncertainty. Empirical data, however, has tended to support a negative, rather than positive, relationship between uncertainty avoidance and innovation adoption. In an analysis of data from forty-five countries, for example, despite the researcher's hypothesis of a positive relationship, uncertainty avoidance was negatively associated with Internet use (Park 2001). In conclusion from this finding, the author suggested that the unforeseen changes that new technology may bring about, and the uncertainty that these create, may have greater weight in influencing individuals' openness to innovations vis-à-vis the uncertainty-mitigating effects that some technologies may have.

A second dimension in Hofstede's taxonomy is that of power distance, reflecting the degree to which the weaker segments of society accept and expect an unequal distribution of power and wealth. Power distance reflects the degree to which members of society are sensitive to status differences and are motivated by the need to conform to their status group or the group with which they identify (Roth 1995). In high power distance cultures, individuals' position in society becomes particularly meaningful, with role assignments and functions being especially restrictive. In

Hofstede's studies, high power distance countries included, for example, the Philippines and Mexico, whereas those low in power distance were the United States and Israel.

Arguments for both positive and negative relationships between power distance and diffusion of innovation have been proposed. Those arguing for a positive relationship suggest that when power distance is high, those in power more readily seek to adopt innovations that help reflect and maintain their status, and the less powerful defer to those in power and are strongly influenced by opinion leaders (Dwyer, Mesak, and Hsu 2005). Contrarily, those arguing for a negative relationship suggest that in low power distance societies, overall communication is better, with fewer barriers, which may elicit the faster adoption of new products (e.g., Chandrasekaran and Tellis 2008). Although some empirical support has been established for the positive relationship (Dwyer, Mesak, and Hsu 2005), most of the studies we have come across provided evidence for a negative relationship, indicating greater resistance to innovation as societal power distance increased (see references in Table 4.1).

Perhaps the most familiar of Hofstede's cultural dimensions is that of individualism versus collectivism. Individualism involves the degree to which individuals are autonomous rather than integrated into groups (Hofstede 2001). Individualistic societies are characterized by loose ties among their members. Examples of individualistic countries in Hofstede's studies are the United States, Great Britain, and Italy, whereas collectivist countries include Japan, Pakistan, and South Korea. These latter societies tend to emphasize loyalty, solidarity, and identification with the group. Some have suggested that the individualistic context facilitates innovative behavior, as in the pursuit of innovations, which involves a tendency to initiate new behaviors, independently of others (Steenkamp, Hofstede, and Wedel 1999). It has therefore been proposed that individualism will be positively associated with the adoption of innovation, whereas collectivism with greater resistance to innovations. Correspondingly, others, proposing a similar link, based their arguments on the particular nature of the innovations studied (e.g., Park 2001; Stremersch and Lemmens 2009). For example, in Stremersch and Lemmens's study of new pharmaceuticals, individualism was expected to correlate with country-level sales of new drugs because of individualistic societies' emphasis on individuals' well-being. Most of the research on individualism and innovation adoption has indeed established a positive relationship between the two (see table 4.1).

Some exceptions, however, have been found. For example, in one study of the diffusion of cellular phones, a negative relationship was found (Stump, Gong, and Chelariu 2010). To explain their finding, the authors proposed that cellular phones serve to reinforce existing ties and thus encourage individuals in collectivist societies to adopt them more than within individualistic societies. It therefore seems that beyond an overall predisposition that individuals in collectivist societies may have against innovations, the specific pattern of relationship between this cultural dimension and innovation adoption will depend on the specific nature of the innovation.

Several of the studies above that explored individualism also looked into Hofstede's masculinity-femininity dimension. Cultural masculinity is defined as the extent to which a society's value system is characterized by male or female attributes (Hofstede 2001). Masculine societies are said to put a greater emphasis on assertiveness, whereas feminine societies set a greater value on nurturance. Specifically, whereas masculine societies place a greater emphasis on ambition and material things feminine societies emphasize helping others and preserving equality (Steenkamp, Hofstede, and Wedel 1999). Results of diffusion studies that considered masculinity are mixed. Positive relationships have been explained by suggesting that masculine societies' materialistic tendencies may be expressed in a greater propensity to purchase new items (Singh 2006; Steenkamp, Hofstede, and Wedel 1999) whereas negative relationships have been typically explained by the nature of the specific innovations studied. For example, feminine societies' emphasis on nurturance was used to explain the negative relationships found between masculinity and new drug sales (Stremersch and Lemmens 2009). Similarly, the societies' emphasis on equality and solidarity was used to explain the (unexpected) negative relationship found between masculinity and the penetration of ERP systems, designed for sharing information within companies and among them (Van Everdingen and Waarts 2003).

The fifth, and most recent, dimension to have been proposed by Hofstede is that of long-term versus short-term orientation. Being introduced twenty-one years following Hofstede's original taxonomy (Hofstede 1980), fewer studies have incorporated this dimension into their research models. The definition of this dimension provides substance for hypothesizing relationships in both directions with innovation adoption. *Long-term societies* are characterized by their emphasis on the values of thrift, perseverance, and a look to the future, as well as being patient and refraining from conspicuous consumption. Contrarily, *short-term societies* place greater value on respect for tradition and stability as well as on sensitivity to social trends in consumption. Indeed, whereas some researchers hypothesized and found positive relationships between long-term orientation and innovation adoption (Van Everdingen and Waarts 2003), others expected and found a negative relationship (Dwyer, Mesak, and Hsu 2005; Stremersch and Lemmens 2009). Clearly, additional research using this dimension should be conducted for this relationship to be better understood in the context of innovation adoption.

Although the studies above provide numerous insights into the country-level contextual factors that explain adoption-of versus resistance-to innovations, some have criticized this literature for not providing a clear and consistent enough picture of the relationships between contextual variables and adoption (Dekimpe, Parker, and Sarvary 2000). Whereas it is relatively clear that individuals in poorer and more conservative countries exhibit greater resistance to innovation, the picture with respect to other societal and cultural factors is not as clear. The most consistent finding for culture was the positive relationship between uncertainty avoidance and resistance to innovation. The relationship for the other cultural variables was much more a function of the specific innovation studied. Related to this point, it has also been noted that only a limited range of innovations has been considered, with a heavy focus on consumer durables, and the majority of studies have relied on data from European countries (Dekimpe, Parker, and Sarvary 2000).

Some researchers explicitly discussed the differences in effects found for economic versus cultural variables, sometimes concluding that culture has a stronger effect than economic variables (Tellis, Stremersch, and Yin 2003), and other times concluding the opposite (Stremersch and Tellis 2004). Stremersch and Tellis (2004) reconcile these different conclusions by highlighting that the dominance of cultural effects was seen when product takeoff was the outcome variable (Tellis, Stremersch, and Yin 2003), whereas the dominance of economic effects was established when the outcome was product sales' growth (Stremersch and Tellis 2004). Stremersch and Tellis go on to suggest that whereas Innovators (Rogers 2003) who adopt before takeoff are driven by cultural factors, early adopters and the early majority, which more substantially contribute to products' sales growth, are more likely driven by affordability concerns. This suggests that resistance would be more effectively predicted by countries' economic standing than by the cultural values they emphasize. Certainly, however, additional research should be conducted to more explicitly compare the impact of economic and cultural factors in explaining resistance to innovation. Furthermore, as the distinction that Stremersch and Tellis exemplifies, more attention should be paid to how variables are operationalized, given that in the current state of affairs, the different operationalizations makes it difficult to compare and generalize findings (Dekimpe, Parker, and Sarvary 2000).

Organizational Contexts and Resistance to Innovation and Change

Beyond the impact of national context, there are many subcontexts within each nation that influence individuals' reactions to change and innovation. As with country-level factors, there are various contextual factors *within* countries that could influence the openness to and ultimate adoption of innovations. Along with previous topics discussed in this book, organizational research on employees' reactions to change provides valuable insights that could help explain resistance to innovation within our present discussion of the role of context. Specifically, several studies considered the impact of the organizational context on employees' resistance to organizational changes. These studies have been reviewed in detail by Oreg, Vakola, and Armenakis (2011) and are summarized in table 4.2.

A number of these studies pertain to the introduction of technological innovations in organizations. One such study considered the implementation of IT systems in eighteen companies involved in government and commercial ventures (Harper and Utley 2001). The researchers were interested in predicting the degree to which the implementation of the IT systems was successful, with a focus on the organizational culture as the predictor. Data were collected from companies from a variety of industries, including companies such as Home Depot, Motorola, Lockheed Martin Energy Systems, and BellSouth Mobility. The success of the IT implementation was assessed from company executives' responses to questions about the IT implementation, including questions about the degree of user involvement, primary causes of failure, and the overall IT payoff. Culture was conceptualized on the basis of O'Reilly, Chatman, and Caldwell's (1991) Organizational Culture Profile, which was reduced to two orthogonal dimensions of *people attributes* and *production attributes*. It was found that the less "people-oriented" the organization's culture was, the less successful was the IT implementation. A more fine grained examination of culture's role indicated that cultural features

TABLE 4.2 **Selected studies of context as a predictor of employees' reactions to organizational change**

Predictor of Reactions to Change	Change Involved	Reference
Supportive and trusting environment	Implementation of TQM at a UK supplier of electrical components	(Coyle-Shapiro and Morrow 2003)
	Reengineering and work redesign at a Canadian hospital	(Cunningham et al. 2002)
	Transition into segmented sales teams at a national sales organization	(Eby et al. 2000)
	Restructuring jobs and technological change at an Australian hospital	(Iverson 1996)
	A variety of changes, including a merger, downsizing, and restructuring at an Aerospace company	(Fugate et al. 2002)
	Continuous change at 3 private-sector and 1 nonprofit organizations	(Madsen et al. 2005)
	A variety of changes including relocation, downsizing and job changes that have to do with the introduction of new technological advances	(Martin et al. 2005)
	Merger of two divisions in a defense-industry organization	(Oreg 2006)
	Relocation of Australian governmental organizations	(Peach et al. 2005)
	Several change including a change in top management, restructuring and downsizing at an Australian public organization	(Rafferty and Griffin 2006)
	Downsizing at an aerospace company	(Spreitzer and Mishra 2002)
	Restructuring and culture change at a number of organizations	(Stanley et al. 2005)
	Radical reorganization at a US government agency	(Wanberg and Banas 2000)
Organizational commitment	Hospital divisional consolidation	(Begley and Czajka 1993)
	Merger of two Fortune 500 companies	(Covin et al. 1996)
	Implementation of TQM at a UK supplier of electrical components	(Coyle-Shapiro 1999)
	Implementation of TQM at a UK supplier of electrical components	(Coyle-Shaipro 2002)
	Implementation of TQM at a UK supplier of electrical components	(Coyle-Shapiro and Morrow 2003)
	Mergers of departments, technological change, and modifications to shift work at a number of hospitals	(Herscovitch and Meyer 2002)
	Changes to university traditions at a public university	(Lau and Woodman 1995)
	Downsizing, and changes to salary and promotion systems at a Korean bank	(Lee and Peccei 2007)
	Continuous change at 3 private-sector and 1 nonprofit organizations	(Madsen et al. 2005)
	Merger, relocation of employees, and job redesign at a number of hospitals	(van Dam 2005)

TABLE 4.2 ***(continued)***

Predictor of Reactions to Change	Change Involved	Reference
Organizational climate and culture	Merger of two UK building societies	(Cartwright and Cooper 1993)
	Reengineering and work redesign at a Canadian hospital	(Cunningham et al. 2002)
	Restructuring jobs and technological change at an Australian hospital	(Iverson 1996)
	Introduction of new end-user IT system at an Australian state government department	(Jones et al. 2005)
	A variety of changes including relocation, downsizing and job changes that have to do with the introduction of new technological advances	(Martin et al. 2005)
	Several changes involved in a merger of two Dutch housing corporations	(van Dam et al. 2008)
Job characteristics	Aircraft manufacturing company transition to 4-day work week	(Bhagat and Chassie 1980)
	Reengineering and work redesign at a Canadian hospital	(Cunningham et al. 2002)
	Transition into segmented sales teams at a national sales organization	(Eby et al. 2000)
	Change in leadership and decentralization of decision making procedures at a hospital	(Hornung and Rousseau 2007)
	Restructuring jobs and technological change at an Australian hospital	(Iverson 1996)
	Organizational change under a new chief at a fire department	(Weber and Weber 2001)

Note: Adapted from tables 3 and 5 in Oreg, Vakola, and Armenakis (2011).

reflecting rigidity, such as rule orientation and compliance, were negatively related to IT implementation success, whereas cultural features that reflect openness, such as autonomy and flexibility, were positively related to it. In other words, the implementation of IT systems in organizations characterized by little supportiveness (low people orientation) and, more specifically, by a rigid atmosphere, was more likely to fail compared with its implementation in supportive and flexible organizational cultures.

Similar findings were obtained in two other studies that involved the introduction of new technology in organization (Jones, Jimmieson, and Griffiths 2005; Martin, Jones, and Callan 2005). In the first study, of two samples of employees from large public organizations, the psychological climate predicted employees' approach toward a multifaceted change, which included changes in technologically related work practices (Martin, Jones, and Callan 2005). Specifically, a climate that was perceived as supportive and friendly was associated with positive appraisals of the organizational change, which in turn predicted better psychological adjustment to the change.

Another study, of smaller scale, involved the introduction of an end-user computing system in a state government department in Queensland, Australia (Jones, Jimmieson, and Griffiths 2005). Following system implementation, employees would be able to access information and submit job-related queries through the new system. The researchers' aim was to predict the degree to which the new system would be used. As hypothesized, employees who characterized their division as including strong human relations values, open communication, participative decision making, and a sense of cohesion, reported higher levels of readiness for the change and an increase in the degree to which the system was used in comparison to employees who had different characterizations of their division. A state of *readiness for change* (Armenakis, Harris, and Mossholder 1993) appears to be key in explaining individuals' openness to innovation, with low readiness increasing the likelihood of resistance.

Overall, research has shown that changes, including those that involve the introduction of innovations, are better accepted in organizations when the organizational environment is supportive and collegial, and when employees have been committed to their organization. Furthermore, specific characteristics of the job, including the degree of autonomy employees have, have also been shown to influence employees' acceptance of, versus resistance to, change (see table 4.2).

Linking Context to Process and Perceived Threats

Each of the above studies, in both the organizational and consumer contexts, demonstrate meaningful relationships between the context within which an innovation is introduced and its degree of acceptance. This relationship is indicated in the direct relationship drawn between context and resistance to innovation in figure 4.1. Hardly any research has been conducted, however, to demonstrate the mechanisms that explain these relationships. How exactly does the context influence adoption intentions? To answer this question, the change process (see chapter 3) has been highlighted as a possible explanation for the context-resistance relationship (van Dam, Oreg, and Schyns 2008). In the van Dam et al. study, the authors examined a merger of two Dutch housing corporations, which involved drastic changes in working procedures and management practices. Employees reported experiencing a great degree of uncertainty, with much variance in the degree to which they resisted the change.

The exogenous predictors in their model involved two aspects of the organizational context. First, in line with the literature reviewed above, the degree to which employees had an open and trusting relationship with their manager (termed *Leader Member Exchange*, or *LMX*) was expected to correlate negatively with employees' resistance to the change. In addition, the degree to which the organization was perceived as fostering continuous development, learning, and growth, termed *development climate*, was also expected to correlate negatively with employees' resistance to change. The explanation for these hypothesized relationships was that supportive and development-oriented organizations should be more likely to exhibit a participative change implementation process involving factors such as the provision of information about the change, opportunities for participation in its implementation, and faith in management's ability to handle the change. These features of the change process are in turn expected to decrease resistance to the change (as discussed in the previous chapter). In other words, the change process was said to mediate the relationship between the change context and resistance to the change. In addition, it was hypothesized that individual characteristics will also predict the level of resistance to change (see chapter 1). The individual characteristics considered were an overall *openness to job changes* (a correlate of dispositional resistance to change) and employees' perceived ability to carry out a broader set of work tasks that go beyond prescribed requirements (termed *role-breadth self-efficacy*), which was relevant for the particular change studied. The research model that was tested is depicted in figure 4.2.

Data were collected from 235 employees, working at a variety of departments, such as technical, secretarial, and customer service. Approximately 16% of the individuals in the sample held a supervisory position. The research model was tested with a Structural Equations Modeling (SEM) analysis, with maximum likelihood estimation. In table 4.3 we present the means, standard deviations, reliabilities, and intercorrelations of the study variables.

The SEM analysis yielded overall support for the research model. Model fit was good (χ^2 [df = 11, N = 235] = 18.58, p = 0.07; χ^2 / df = 1.69; CFI = 0.98; RMSEA = 0.05; AIC = 86.58), and all of the hypothesized paths were significant, with the exception of the individual characteristic

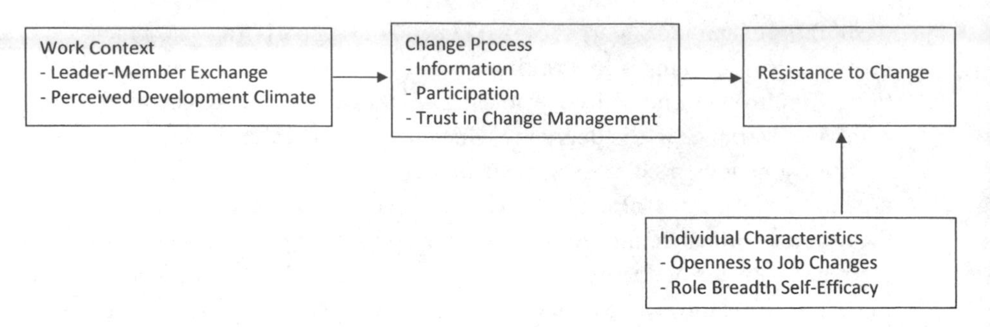

FIGURE 4.2. Research model in van Dam, Oreg, and Schyns (2008).

of role breadth self-efficacy, which was not significantly correlated with resistance to change. The standardized regression coefficients from the analysis are presented in table 4.4.

Thus, in this study, the change process mediated the relationships of LMX and perceived development climate with resistance to change. Employees whose work context was such that they experienced a trusting relationship with their manager, and a strong development climate in the organization, reported receiving more information, opportunities for participation, and trust in the change management, which subsequently led them to exhibit less resistance to the change. The study highlights at least one mechanism—the change implementation process—through which context influences resistance to change. This relationship is also indicated in the mediating role of process in figure 4.1. In the broader consumer context, we can say that differences in national cultures often yield differences in the manner and style through which innovations are introduced, which in turn yield different reactions to the innovation across nations. Despite its conceptual appeal, we are not aware of any research that directly tested this possibility.

The examples opening this chapter bring us to suggest yet another mechanism through which context may influence resistance to change—the perceived utility/threat. It may well be that the context within which changes and innovations are introduced influences the utility or threat individuals perceive in the change or innovation. Starbucks was perceived in Israel as having very low utility, given the well-established coffee culture that existed in Israel prior to Starbucks' venture in the country. Similarly, Wal-Mart failed in Germany because, in the context of the competitive German market, and German values and priorities, German consumers

TABLE 4.3 **Means, standard deviations, and intercorrelations**

	Variable	*M*	*SD*	1	2	3	4	5	6	7	8
1	LMX	3.48	0.31								
2	Development climate	3.12	0.58	0.61***							
3	Information	3.34	0.34	0.39***	0.38***						
4	Participation	2.06	0.93	0.31***	0.41***	0.44***					
5	Trust in management	3.35	0.33	0.50***	0.48***	0.51***	0.38***				
6	Openness to job changes	3.19	0.76	−0.05	−0.07	0.08	0.04	0.06			
7	Role breadth self efficacy	3.32	0.56	−0.04	0.08	0.06	0.13*	0.02	0.15*		
8	Tenure	10.83	9.39	−0.03	−0.06	−0.06	0.02	−0.03	−0.18**	−0.06	
9	Resistance to change	2.75	0.73	−0.38***	−0.42***	0.48***	−0.49***	−0.43***	−0.23***	−0.14*	0.19**

Note: $N = 235$. Adapted from van Dam, Oreg, and Schyns (2008).
*$p < 0.05$.
**$p < 0.01$.
***$p < 0.001$.

TABLE 4.4 **Estimated regression coefficients from the SEM structural model**

	Dependent variables			
	Information	Participation	Trust in change management	Resistance to change
Change context				
Leader-Member exchange	0.24 ***	0.19 **	0.33 ***	
Development climate	0.24 ***	0.36 ***	0.28 ***	
Change process				
Information				−0.23 ***
Participation				−0.31 ***
Trust in management				−0.19 ***
Individual characteristics				
Openness to job changes				−0.16 **
Role breadth self-efficacy				−0.05
Job tenure (control variable)				0.15 **

Note: Adapted from van Dam et al. (2008); $N = 235$.
*$p < 0.05$.
**$p < 0.01$.
***$p < 0.001$.

simply perceived Wal-Mart as having very little to offer. Their innovations failed because they weren't perceived as valuable.

As in the previous chapter, here too we could use Oreg's (2006) data to test this possibility. Let us recall the study of the organizational restructuring, discussed in the previous chapter. Among the factors that predicted employees' resistance to the change was their belief that the change threatens their power and prestige in the organization. Another factor, shown in the 2006 study to strongly predict resistance to change, was a lack of trust in management. Indeed, as discussed above, one of the contexts that has been repeatedly shown to predict resistance to change and innovation in organizations is the lack of a supportive and trusting, and trustworthy, environment. This latter conceptualization of trust involves an overall sense of faith in the organization's management, even prior to the implementation of any given change that serves and the organizational context within which the change was implemented.

Thus in the 2006 study both lack of trust and perceived threats to power and prestige were hypothesized to directly relate to employees' resistance to the change. Given our discussion above, we propose that at least some of the impact that trust, as an aspect of the organizational context, has on resistance to change may derive from the impact that a lack of trust has on employees' perceived threats. In other words, perceived threats may mediate the relationship between lack of trust and resistance to change. Oreg's 2006 data allowed us to test the proposed mediation effect.

Here too, mediation was tested using Baron and Kenny's (1986) steps (see table 4.5). As can be seen in table 4.5, threats to power and prestige partially mediated the relationship between lack of trust and resistance to change. Although the effect of "lack of trust" on resistance remains significant when controlling for perceived threats, the effect is nevertheless weaker than in the noncontrolled analysis. This suggests that context has both direct and indirect effects on resistance. The indirect effect was demonstrated in this study through the mediating role of perceived threats. In the previous analysis, from van Dam and colleagues (2008), we demonstrated the indirect effect through the mediating role of the change process.

Together with the mediating effect that perceived threats have on the relationship between innovation process and resistance, these findings are all portrayed in figure 4.1. In addition to direct effects that exist between each of the predictors in the model and resistance to innovation, our findings suggest that the innovation context influences the innovation process, which in turn influences the perceived threats, which can ultimately elicit resistance to innovation.

TABLE 4.5 **Steps in testing perceived threat as a mediator of the trust-resistance relationship**

	B	Std. Err	β	R^2
First regression: (Resistance to change)				0.09
Lack of trust in management	0.26	0.06	0.30***	
Second regression: (Threats to power and prestige)				0.11
Lack of trust in management	0.18	0.04	0.33***	
Third regression: (Resistance to change)				0.14
Lack of trust in management	0.19	0.06	0.23**	
Threats to power and prestige	0.36	0.12	0.23**	

Note: Dependent variables are in parentheses.
**$p < 0.01$.
***$p < 0.001$.

Conclusions

Compared with the other situational (chapters 2 and 3) antecedents of resistance that we discussed in this part of the book, the innovation's context is probably the least malleable and least subject to marketers' control. Whereas marketers can control the innovation process, and can even influence the degree to which consumers will perceive an innovation as threatening, most aspects of the change context are far beyond their control. Obviously, marketers cannot influence the cultural, societal, or economic settings within which their products are introduced. But they can, and often do, take into consideration contextual factors when deciding where to introduce their products to begin with. We believe that one of the benefits of our review is the structure it provides by mapping out the various types of contextual variables that have been considered as antecedents of innovation adoption and rejection.

From a practical perspective, the list of context categories included in table 4.1 can be used as a checklist when preparing to launch an innovation in a new context. The timing of an innovation's introduction vis-à-vis its introduction in other countries, and the various variables within the societal, economic, and cultural factors we discussed, can each be reviewed to determine the likelihood of resistance within the new proposed context. Furthermore, although one cannot influence the change context, awareness of the links we propose between innovation context, process, and perceived threat can help marketers prepare for, and in a sense shape, those factors that *can* be dealt with to mitigate resistance.

Indeed, one of our aims in this chapter was to demonstrate the role of the contextual factors vis-à-vis other antecedents of resistance to innovation. By proposing and testing mechanisms that could explain how context ultimately comes to influence resistance, we hope to have provided a more comprehensive understanding of the resistance process. Each of the resistance antecedents we discussed has its role in explaining resistance, yet as our analyses indicate, they vary in the immediacy of their impact on consumers' ultimate reaction to the innovation.

As the model outlined in figure 4.1 suggests, consumer characteristics, and the threats that consumers may perceive in the innovation, are the most proximal predictors of resistance. Their influence on resistance is the most immediate and direct. Not surprisingly, these factors are also among the most common explanations given for consumers' rejection of products. Somewhat less intuitive, however, are the more distal factors of innovation process and context. Acknowledging the role that the innovation context has in shaping both the innovation process and the perceived utility versus threat, and the role that process has in shaping consumers' perceived utility/threat, provides for a better understanding of what it is about the innovation's context and process that yields resistance. It more concretely and explicitly illustrates how the setting and manner in which innovations are introduced are translated into consumers' perceptions of the product. This information can then be used to design the innovation's marketing process, and to circumvent consumers' potential perceived threats, in order to reduce the resistance that the given context predisposes consumers to experience. For example, marketers who consider launching an innovation in a country high on uncertainty avoidance, should know to give particular attention to consumer participation during the innovation's launch and to provide detailed information about the innovation that could preempt consumers' perceived threats and thus alleviate their resistance.

So far in this book, our focus has been on understanding where resistance comes from. What are the factors that increase its likelihood, and consequently, what may be done to alleviate or circumvent it. Our focus has been primarily on the individual's perspective given that all forms and manifestations of resistance begin with the resistance of the individual. Obviously, however, what is of most interest to marketers is the amalgamation of individuals' resistancc that forms to form market-level resistance. It is this type of resistance, and the processes through which it may develop, on which we focus in the following chapters. We will begin by linking individual-level and market-level resistances. We demonstrate such a linkage in the next chapter by showing the market-level impact of alleviating individuals' resistance. Assuming that understanding the sources of consumers' resistance can help us alleviate at least part of it, and thus expedite consumers' adoption of products, we demonstrate in chapter 5 the substantial financial impact that such expedited adoption could have.

References

Armenakis, Achilles A., Stanley G. Harris, and Kevin W. Mossholder. 1993. "Creating Readiness for Organizational Change." *Human Relations* 46: 681–703.

Baron, Reuben M., and David A. Kenny. 1986. "The Moderator-Mediator Variable Distinction in Social Psychological Research: Conceptual, Strategic, and Statistical Considerations." *Journal of Personality and Social Psychology* 51 (6): 1173–82.

Basker, E. 2007. "The Causes and Consequences of Wal-Mart's Growth." *Journal of Economic Perspectives* 21 (3): 177–98.

Begley, Thomas M., and Joseph M. Czajka. 1993. "Panel Analysis of the Moderating Effects of Commitment on Job Satisfaction, Intent to Quit, and Health Following Organizational Change." *Journal of Applied Psychology* 78 (4): 552–56.

Berry, L. L., V. Shankar, J. T. Parish, S. Cadwallader, and T. Dotzel. 2006. "Creating New Markets through Service Innovation." *MIT Sloan Management Review* 47 (2): 56.

Bhagat, Rabi S., and Marilyn B. Chassie. 1980. "Effects of Changes in Job Characteristics on Some Theory-Specific Attitudinal Outcomes: Results from a Naturally Occurring Quasi-Experiment." *Human Relations* 33 (5): 297–313.

Cartwright, Sue, and Cary L. Cooper. 1993. "The Psychological Impact of Merger and Acquisition on the Individual: A Study of Building Society Managers." *Human Relations* 46 (3): 327.

Chandrasekaran, Deepa, and Gerard J. Tellis. 2008. "Global Takeoff of New Products: Culture, Wealth, or Vanishing Differences?" *Marketing Science* 37 (5): 844–60.

Chavan, A. L., D. Gorney, B. Prabhu, and S. Arora. 2009. "The Washing Machine That Ate My Sari: Mistakes in Cross-Cultural Design" *Interactions* 16 (1): 26–31.

Christopherson, S. 2007. "Barriers to 'US Style' Lean Retailing: The Case of Wal-Mart's Failure in Germany." *Journal of Economic Geography* 7 (4): 451.

Coussin, Orna. 2003. "Coffee to Go: Why Aren't America's Finest Winning over the Israeli Market?" In Ha'aretz.

Covin, Teresa Joyce, Kevin W. Sightler, Thomas A. Kolenko, and Keith R. Tudor. 1996. "An Investigation of Post-Acquisition Satisfaction with the Merger." *Journal of Applied Behavioral Science* 32 (2): 125–42.

Coyle-Shapiro, Jacqueline A. M. 1999. "Employee Participation and Assessment of an Organizational Change Intervention: A Three-Wave Study of Total Quality Management." *Journal of Applied Behavioral Science* 35 (4): 439–56.

———. 2002. "Changing Employee Attitudes: The Independent Effects of TQM and Profit Sharing on Continuous Improvement Orientation." *Journal of Applied Behavioral Science* 38 (1): 57–77.

Coyle-Shapiro, Jacqueline A. M., and Paula C. Morrow. 2003. "The Role of Individual Differences in Employee Adoption of TQM Orientation." *Journal of Vocational Behavior* 62 (2): 320–40.

Cunningham, Charles E., Christel A. Woodward, Harry S. Shannon, John MacIntosh, Bonnie Lendrum, David Rosenbloom, and Judy Brown. 2002. "Readiness for Organizational Change: A Longitudinal Study of Workplace, Psychological, and Behavioural Correlates." *Journal of Occupational and Organizational Psychology* 75 (4): 377–92.

Dekimpe, M., P. M. Parker, and M. Sarvary. 2000. "Multimarket and Global Diffusion." In *New-Product Diffusion Models*, edited by Vijay Mahajan, Eitan Muller, and Yoram Wind. New York: Springer.

Dwyer, S., H. Mesak, and M. Hsu. 2005. "An Exploratory Examination of the Influence of National Culture on Cross-National Product Diffusion." *Journal of International Marketing* 13 (2): 1–27.

Eby, Lillian T., Danielle M. Adams, Joyce E. A. Russell, and Stephen H. Gaby. 2000. "Perceptions of Organizational Readiness for Change: Factors Related to Employees' Reactions to the Implementation of Team-Based Selling." *Human Relations* 53 (3): 419–42.

Fugate, Mel, Angelo J. Kinicki, and L. Christine Scheck. 2002. "Coping with an Organizational Merger over Four Stages." *Personnel Psychology* 55 (4): 905–28.

Gatignon, H., J. Eliashberg, and T. S. Robertson. 1989. "Modeling Multinational Diffusion Patterns: An Efficient Methodology." *Marketing Science* 8 (3): 231–47.

Haapaniemi, T. P., and S. J. Mäkinen. 2008. "Moderating Effect of National Attributes and the Role of Cultural Dimensions in Technology Adoption Takeoff." *Management Research News* 32 (1): 5–25.

Haig, Matt. 2005. *Brand Failures: The Truth about the 100 Biggest Branding Mistakes of All Times*. London: Kogan Page.

Harper, G. R., and D. R. Utley. 2001. "Organizational Culture and Successful Information Technology Implementation." *Engineering Management Journal* 13 (2): 11–15.

Helsen, K., K. Jedidi, and W. S. DeSarbo. 1993. "A New Approach to Country Segmentation Utilizing Multinational Diffusion Patterns." *Journal of Marketing* 57 (4): 60–71.

Herscovitch, Lynne, and John P. Meyer. 2002. "Commitment to Organizational Change: Extension of a Three-Component Model." *Journal of Applied Psychology* 87 (3): 474–87.

Hofstede, Geert. 1980. *Culture's Consequences, International Differences in Work-Related Values*. Beverly Hills, CA: Sage.

———. 2001. *Culture's Consequences: Comparing Values, Behaviors, Institutions, and Organizations across Nations*. 2nd ed. Thousand Oaks, CA: Sage.

Hornung, Severin, and Denise M. Rousseau. 2007. "Active on the Job—Proactive in Change: How Autonomy at Work Contributes to Employee Support for Organizational Change." *Journal of Applied Behavioral Science* 43 (4): 401–26.

Iverson, R. 1996. "Employee Acceptance of Organizational Change: The Role of Organizational Commitment." *International Journal of Human Resource Management* 7 (1): 122–49.

Jones, Renae A., Nerina L. Jimmieson, and Andrew Griffiths. 2005. "The Impact of Organizational Culture and Reshaping Capabilities on Change Implementation Success: The Mediating Role of Readiness for Change." *Journal of Management Studies* 42: 361–86.

Knorr, A., and A. Arndt. 2003. "Why Did Wal-Mart Fail in Germany?" In *Materialien des Wissenschaftsschwerpunktes "Globalisierung der Weltwirtschaft,"* edited by A. Knorr, A. Lemper, A. Sell, and K. Wohlmuth. Vol. 24. Institut für Weltwirtschaft und Internationales Management, Universität Bremen.

Kumar, V., J. Ganesh, and R. Echambadi. 1998. "Cross-National Diffusion Research: What Do We Know and How Certain Are We?" *Journal of Product Innovation Management* 15 (3): 255–68.

Landler, M. 2006. "Wal-Mart Decides to Pull Out of Germany." *New York Times*, July 28.

Lau, Chung-Ming, and Richard W. Woodman. 1995. "Understanding Organizational Change: A Schematic Perspective." *Academy of Management Journal* 38 (2): 537–54.

Lee, Jaewon, and Riccardo Peccei. 2007 "Perceived Organizational Support and Affective Commitment: The Mediating Role of Organization-Based Self-Esteem in the Context of Job Insecurity." *Journal of Organizational Behavior* 28 (6): 661–85.

Madsen, Susan R., Duane Miller, and Cameron R. John. 2005. "Readiness for Organizational Change: Do Organizational Commitment and Social Relationships in the Workplace Make a Difference?" *Human Resource Development Quarterly* 16 (2): 213–33.

Martin, Angela J., Elizabeth S. Jones, and Victor J. Callan. 2005. "The Role of Psychological Climate in Facilitating Employee Adjustment during Organizational Change." *European Journal of Work and Organizational Psychology* 14 (3): 263–83.

Mirza, Faryal, 2010. "Americans Fail to Get a Taste for Rivella." Accessed November 22, 2010, available at http://www.swissinfo.ch/eng/Americans_fail_to_get_a_taste_for_Rivella.html?cid=4802966.

Nilsson, D. 2007. "A Cross-Cultural Comparison of Self-Service Technology Use." *European Journal of Marketing* 41 (3/4): 367–81.

Oreg, Shaul. 2006 "Personality, Context, and Resistance to Organizational Change." *European Journal of Work and Organizational Psychology* 15: 73–101.

Oreg, Shaul, Maria Vakola, and Achilles A. Armenakis. 2011. "Change Recipients' Reactions to Organizational Change: A Sixty-Year Review of Quantitative Studies." *Journal of Applied Behavioral Science* 47 (4): 461–524.

O'Reilly, Charles A., Jennifer Chatman, and David F. Caldwell. 1991. "People and Organizational Culture: A Profile Comparison Approach to Assessing Person-Organization Fit." *Academy of Management Journal* 34 (3): 487–516.

Park, H. 2001. "Cultural Impact on Internet Connectivity and Its Implication." *Journal of Euromarketing* 10 (3): 5–22.

Patterson, P. G., J. Scott, and M. D. Uncles. 2010. "How the Local Competition Defeated a Global Brand: The Case of Starbucks." *Australasian Marketing Journal (AMJ)* 18 (1): 41–47.

Peach, Megan, Nerina L. Jimmieson, and Katherine M. White. 2005. "Beliefs Underlying Employee Readiness to Support a Building Relocation: A Theory of Planned Behavior Perspective." *Organization Development Journal* 23 (3): 9–22.

Peres, R., E. Muller, and V. Mahajan. 2010. "Innovation Diffusion and New Product Growth Models: A Critical Review and Research Directions." *International Journal of Research in Marketing* 27 (2): 91–106.

Png, I. P. L., B. C. Y. Tan, and K. L. Wee. 2002. "Dimensions of National Culture and Corporate Adoption of IT Infrastructure." *Engineering Management, IEEE Transactions On* 48 (1): 36–45.

Rafferty, Alannah E., and Mark A. Griffin. 2006. "Perceptions of Organizational Change: A Stress and Coping Perspective." *Journal of Applied Psychology* 91 (5): 1154–62.

Rogers, Everett M. 2003. *Diffusion of Innovations*. 5th ed. New York: Free Press.

Roth, M. S. 1995. "The Effects of Culture and Socioeconomics on the Performance of Global Brand Image Strategies." *Journal of Marketing Research* 32 (2): 163–75.

Shaked, Ra'anan. 2003. "For America, the Coffee Virgin, Starbucks Is the First Man and the Perfect Lover. For Us [in Israel], It Is No More Than a Pimply Boy, Trying to Interest Us with His Dubious Graces [in Hebrew]." Ynet: Yediot Aharonot.

Singh, S. 2006. "Cultural Differences In, and Influences On, Consumers' Propensity to Adopt Innovations." *International Marketing Review* 23 (2): 173–91.

Spreitzer, Gretchen M., and Aneil K. Mishra. 2002. "To Stay or to Go: Voluntary Survivor Turnover Following an Organizational Downsizing." *Journal of Organizational Behavior* 23 (6): 707–29.

Stanley, David J., John P. Meyer, and Laryssa Topolnytsky. 2005. "Employee Cynicism and Resistance to Organizational Change." *Journal of Business and Psychology* 19 (4): 429–59.

Steenkamp, Jbem, F. Hofstede, and M. Wedel. 1999. "A Cross-National Investigation into the Individual and National Cultural Antecedents of Consumer Innovativeness." *Journal of Marketing* 63 (2): 55–69.

Stremersch, S., and A. Lemmens. 2009. "Sales Growth of New Pharmaceuticals across the Globe: The Role of Regulatory Regimes." *Marketing Science* 28 (4): 690–708.

Stremersch, S., and G. J. Tellis. 2004. "Understanding and Managing International Growth of New Products." *International Journal of Research in Marketing* 21 (4): 421–38.

Stump, R. L., W. Gong, and C. Chelariu. 2010. "National Culture and National Adoption and Use of Mobile Telephony." *International Journal of Electronic Business* 8 (4): 433–55.

Takada, H., and D. Jain. 1991. "Cross-National Analysis of Diffusion of Consumer Durable Goods in Pacific Rim Countries." *Journal of Marketing* 55 (2): 48–54.

Tellis, G. J., S. Stremersch, and E. Yin. 2003. "The International Takeoff of New

Products: The Role of Economics, Culture, and Country Innovativeness." *Marketing Science* 22 (2): 188–208.

van Dam, Karen. 2005. "Employee Attitudes toward Job Changes: An Application and Extension of Rusbult and Farrell's Investment Model." *Journal of Occupational and Organizational Psychology* 78 (2): 253–72.

van Dam, Karen, Shaul Oreg, and Birgit Schyns. 2008. "Daily Work Contexts and Resistance to Organizational Change: The Role of Leader-Member Exchange, Perceived Development Climate, and Change Process Quality." *Applied Psychology: An International Review* 57 (2): 313–34.

Van Everdingen, Y. M., and E. Waarts. 2003. "The Effect of National Culture on the Adoption of Innovations." *Marketing Letters* 14 (3): 217–32.

Wanberg, Connie R., and Joseph T. Banas. 2000. "Predictors and Outcomes of Openness to Changes in a Reorganizing Qorkplace." *Journal of Applied Psychology* 85 (1): 132–42.

Weber, Paula S., and James E. Weber. 2001. "Changes in Employee Perceptions during Organizational Change." *Leadership and Organization Development Journal* 22 (5/6): 291–300.

Zinkhan, G., A. E. Thyroff, A. Remple, and H. Kim. 2009. "Adding Goods and Services." In *Business Fundamentals*, edited by D. J. McCubbrey. Global Text Project.

PART II
Resistance Manifestations

CHAPTER FIVE

Lagging: Innovation in Disguise

This chapter is based on the article: Goldenberg, J., and S. Oreg. 2007. "Laggards in Disguise: Resistance to Adopt and the Leapfrogging Effect."—*Technological Forecasting and Social Change* 74: 1272–81.

Introduction

In late July 2014, two men entered the Fifth Avenue Apple store in New York City. While it was clear that they were close friends, their behavior in the store was markedly different. David had a spark in his eyes; he seemed to be well acquainted with some of the store's employees and was familiar with all of the products and gadgets in the store. He discussed with the staff several improvements he had in mind for some of the products, had suggestions for how to fix some of their known bugs, and practically begged them to reveal what the next innovation to arrive in the store would be. Beyond being a consumer, David behaved almost like a company employee, in charge of product development. He is what has been clearly identified in marketing as an *innovator*.

Aaron's mannerisms, however, clearly revealed that he didn't want to be there. He was only there because he and David were on their way to a meeting and had some time to kill before it started. Aaron couldn't care less about the devices in the store and frequently glanced at his watch, literally counting the seconds. Despite what you may think at this point, Aaron has a smartphone, and knows everything he needs to know about computers. He simply has no interest in innovations per se. He is not an innovator, and in marketing is considered a member of the *majority*. Aaron belongs to the most important and largest segment among innovation adopters.

As we will explain below, the term *Innovator* encompasses a number of features, not all of which are interchangeable. For now, we focus on perhaps the key feature of Innovators, which is their choice to adopt early. A vast amount of attention has been given to early adopters such as David. Firms need them, marketers love them, and they are typically respected and even envied. You can see an admiring crowd congregate around the person who brought the newest generation smartphone to work, bought the latest glassware (such as Google Glass), or was the first to install Windows 10. People gather around in veneration, as if they were witnessing the discovery of fire, or observing the introduction of the wheel. Without early adopters, firms would have a difficult time surviving, because if it were not for these trailblazers leading the way, there would be no mass market to follow. There are those who would argue that, as consumers, Innovators are not very smart; they buy a very expensive and faulty version of a new, unproven product. Nevertheless, as we further demonstrate in chapter 6, it is clear that they are key to the social process of innovation adoption.

Far less attention has been given to consumers such as Aaron, who are typically assumed to be late to purchase. These individuals' choices are at best ignored and at worst ridiculed. In the extreme, they are called Laggards, and are assumed to be the last to adopt. Their lagging appears to be inherent and so firms expect little of them, having no faith in gaining significant profits from them.

In this chapter, we challenge this view and suggest that it may be premature to give up on this typically late adopting sector. We suggest that while Laggards are slow in switching to a product's new generation, once they *do* switch, they often outstrip most consumers with their purchase. Through what we call the *consumer leapfrogging effect*, we describe how a typical Laggard may come to play the role of an Innovator.

After describing the leapfrogging effect, we move on to discuss its theoretical and practical implications. As we will describe, expediting Laggards' leaps may prove to be quite lucrative. Laggards' resistance could be good news for marketers—addressing them could substantially increase firms' NPV (Net Present Value).

This chapter constitutes the turning point in the two foci we have in the book. In part I, our focus was on resistance at the individual level and the mechanisms that can instigate it. In this part, we shift our focus to several aggregate-level outcomes, or manifestations, of resistance. The most simple manifestation of resistance would be if all, or most, individuals resist the innovation to the degree of avoiding its adoption altogether, thus leading to the innovation's complete failure. Although failures are distressingly common (even if not in such an extreme and simplistic form), they

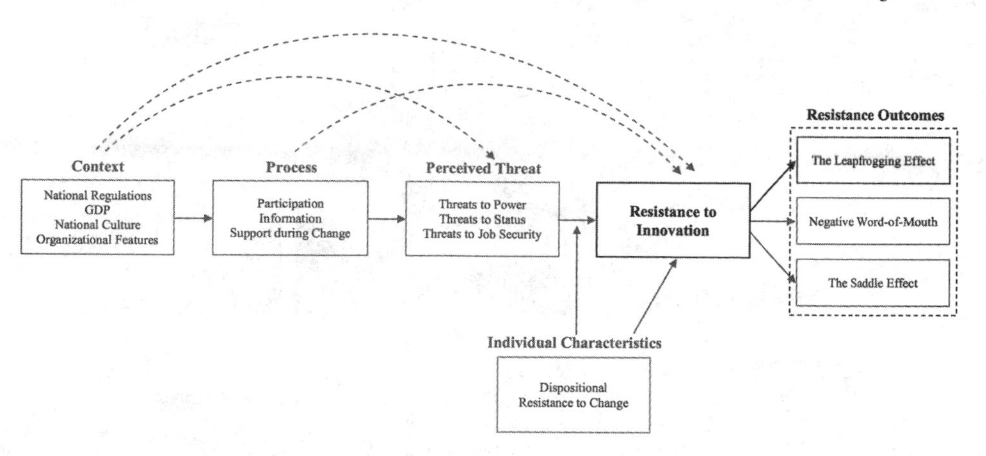

FIGURE 5.1. Resistance and the leapfrogging effect.

represent quite a trivial phenomenon, on which we will not focus. Our focus here will be on those cases in which individuals' behavior is similar to that of Aaron in the example opening this chapter. At the aggregate level, such behavior yields phenomena that, while not a failure, are quite intriguing and meaningful. By understanding such phenomena, firms can either avoid dramatic and costly effects, such as the Saddle (chapter 7), and overall decreases in market size (chapter 6), or even turn resistance on its head, to yield increases in profits (this chapter). We therefore turn now to introduce the first of three aggregate consequences of individuals' resistance: the leapfrogging effect (see figure 5.1).

Laggards

Laggards constitute one of the five segments of innovation adoption styles identified by Rogers (2003). As a group, they have been defined as those who are last to adopt an innovation (following Innovators, Early Adopters, the Early Majority, and the Late Majority). Although, as we discussed in chapter 1, lagging may often result from dispositional resistance, one of the main points we try to make in this chapter is that lagging is not synonymous with being dispositionally resistant. But before explaining this distinction, let us first review what we know about Laggards.

Although some understanding of the Laggard phenomenon can be gained through studies of the broader concept of resistance to change, academic research that has focused directly on Laggards is quite scant. In

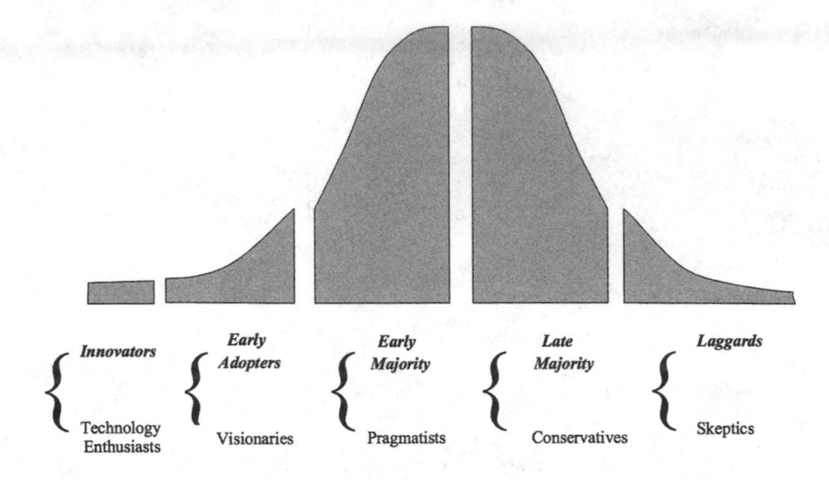

FIGURE 5.2. Rogers's (2003) adopting segments.

figure 5.2, we present a well-known classification of innovation adoption over time.

Because research on late adopters has been so scant, we do not know much about the size of this group. The classification of the five consumer segments in figure 5.1 is based on fixed deviations in a normal distribution. Accordingly, Laggards constitute 16% of the market. Other classifications, based on different principles (such as the adoption slope's last point of deflection; see in chapter 7), are generally consistent, although somewhat smaller, than this assessment (e.g., Mahajan, Muller, and Srivastava 1990).

One of the very few articles focusing on Laggards was penned back in 1970. In this short paper, Kenneth Uhl, Roman Andrus, and Lance Poulsen decried the lack of research on Laggards, pointing out the fact that they are a significant segment of the market and understanding them would help firms meet the needs of all late adopting sectors. Laggards present us with an unadulterated case study of those forces that prevail among a range of market segments. Uhl and colleagues tested the adoption of new brands of food by families in Cedar Rapids, Iowa. Although they emphasized that they wanted to directly assess lagging, they actually used a list of variables derived from earlier studies of Innovators. They discovered that Laggards' family income was lower and brand loyalty was higher than those of other consumers. Whereas Laggards' lower income does not bode well for firms' ability to profit from them, Laggards' great brand loyalty makes them valuable customers once they adopt. Uhl and colleagues also discovered that family size was the best discriminator between Laggards and earlier adopters, with Laggards typically coming from smaller families. Uhl and colleagues speculated that the same characteristics that propel Innovators to adopt new products—newness and originality—might actually discourage Laggards from adopting such products. For Laggards, the new is unproven and risky.

The Uhl et al. study, as well as others that focused on Innovators, found Laggards to be low in incomes, levels of education, social status, and social mobility (Rogers 2003). Such characterizations suggest that lagging is driven by stable individual characteristics, such as those on which we focused in chapter 1. This implies that lagging cannot be changed, which might explain marketers' and researchers' lack of interest in further investigating the lagging phenomenon. We believe that such conclusions are premature, due to what we call the *consumer leapfrogging effect*.

The Consumer Leapfrogging Effect

Consider again Aaron, from the description opening this chapter. Beyond what we have already said about him, Aaron also enjoys listening to music as he jogs every morning. Until a few years ago, Aaron used his 1985 Walkman and was accustomed to cassettes. His entire music collection was recorded on cassettes and he had always preferred using the Walkman rather than switching to newer devices. The fact that his friends all made fun of him made little difference to him. He knows what he likes.

Now, what kind of consumer is Aaron? By definition, Innovators are the first and Laggards are the last to adopt a new product. However, beyond the simple denotations of the term *Laggard*, it also connotes the conservative and backward character of the adopter. Based on the description above, is Aaron likely to be a Laggard? His preference for an antiquated product and his unwillingness, and even resistance, to upgrade would appear to characterize him as one. Indeed, Laggards are often described as "localites" and traditionalists (Rogers 2003). However, do such behavioral inclinations to resist innovation necessarily correspond with late adoption?

We argue that Aaron will not necessarily be a Laggard, and, moreover, that in some situations, he may show up on adoption curves as an

Innovator. As we explain below, when multigenerational products are involved, dispositional resistors (see chapter 1), who are inherently predisposed to hold on to their old products, are more likely to skip product generations, that is, to *leapfrog*. When this happens, someone who possesses the dispositional characteristics of a typical Laggard could very well become an Innovator.

Diffusion models generally look at each product as distinct and independent of other products, and assume that, over time, all members of its potential market will either adopt or decide not to adopt, with nonadopters no longer constituting part of the market. This, however, is often inaccurate. In many cases, products evolve in the form of successive product generations that satisfy the same needs but through an entirely different technology. Each of these product generations can be considered a new product in itself. Music players that came out after the Walkman are a good example. Although completely different from the Walkman in their functions, technology, and quality, CD players and MP3 players address the same fundamental need of listening to music "on the move." In such cases of successive product generations, adoption curves are often incomplete and encompass much less than the entire market.

Imagine now what happened when Aaron finally realized that the sound generated by his Walkman was indeed much poorer than that produced by all of his friends' players, and that the machine was also much bulkier and heavier. Apparently, after so many years, even Aaron acknowledged that it was time to buy something new. In fact, we could even posit that Aaron's Walkman simply broke down. At any rate, the question was: what kind of player would he purchase? Until several years ago, he would have had the option of buying a portable CD player, a MiniDisc, or various versions of the MP3 player.

Let's assume that after discussing the matter with his friends, Aaron purchased the most technologically advanced player. An examination of Aaron's time of purchase, as it appears on the latest product's adoption curve, would identify him as a classic Innovator, among the first to purchase the new product. This type of generation skipping would have made Aaron an Innovator, based on his position on a specific product's life-cycle curve, despite his dispositional tendency to resist innovations. This does not necessarily mean that leaping to the latest technology always results in the adopter appearing as an Innovator, because the "latest" technology may have already been around for a while. Nevertheless, the leaper will still likely appear much earlier on adoption curves than we would have predicted from his or her dispositional tendencies, and on occasion, when the leap happened to occur concurrently with the launch of a new generation, will indeed appear as an Innovator.

Like Aaron, almost all resistors come to a point where they ultimately upgrade their products. What makes them resistors is not that they never upgrade, but that they upgrade *less frequently* than others. One question is how far will they upgrade? Although the newest products may often be more expensive than midrange ones, the more substantial switching costs for Aaron have to do with adapting to the new technology. For Aaron, upgrading meant abandoning the hundreds of cassettes he has accumulated over the years and instead buying CDs or purchasing the rights to download music to his MP3 player. In view of the costs involved in upgrading, it should make little difference to him whether he switches to a portable CD player or the newest device.

Therefore, even though resistors take much longer to upgrade than other consumers, once they upgrade they may very well upgrade to the latest technology available. Contrary to other consumers, however, resistors have held on to their old products long enough for several new generations of products to have joined the market. Thus when resistors upgrade, they will often need to skip several generations to reach the most recent technologies. We call this phenomenon the *consumers' leapfrogging effect.*

A nice example of such leapfrogging was told on an Internet forum about solutions to IT problems. One participant shared his brother's story:

> My brother, a CPA, has his own business, and generally only does an upgrade or buys something when a fire extinguisher gets involved.
>
> He could no longer upgrade the software, and win-98 was getting pretty rank, when something critical finally failed. The replacement was easy enough, but it was incompatible with pretty much everything else in the office.
>
> After 2 weeks of zero productivity for his staff, he relented and gave me his checkbook . . .
>
> I bought all brand new pc's with identical innards, two brand new servers, all brand new drives, raid arrays, upgraded the network to GB ethernet, made a system master disk and blasted it to all the machines, and diagrammed the network. Sadly, the only lesson he got from all that was that the total cost of the complete rebuild was way less than what he would have spent on all of the ongoing upgrades he should have been doing all along.
>
> *sigh*[1]

This CPA, usually a stubborn, classic Laggard, ended up leapfrogging over multiple generations of products after a glitch paralyzed his entire office for two weeks. He hired the services of an expert (his brother) and upgraded not only his software and hardware to the latest generations available, but also restructured the entire infrastructure of his network. Incidentally, this CPA's leapfrogging made financial sense; he spent less money through this spectacular overhaul in one act of leapfrogging than he would have by making periodic updates. As we note throughout the book, resistance has its advantages. This does not prevent his expert brother from adopting a somewhat condescending view, as early adopters and society at large often do, toward Laggards—this in spite of Lagging often making perfect sense.

The multigenerational quality of the products in the CPA's case presented him with the opportunity to leapfrog and become an Innovator. When products do not involve generations, resistors are always among the last to adopt, which by definition makes them Laggards. For these products, then, the terms *Laggard* and *resistor* may be used interchangeably. However, when an industry's products involve several generations (for example, Walkmans, portable CD players, MiniDiscs, and MP3 players), the proclivity for resistance does not necessarily correspond with lagging by its formal definition. Contrary to the accepted definition of Laggards—those who are last to purchase the product—it may be more appropriate to define them in these contexts as those who hold on to their products longest and who are the last to *switch*.

The tendency to lag is therefore not synonymous with being a Laggard. As discussed in chapter 1, research on the concept of resistance to change has uncovered individual differences in people's inclinations to resist changes and innovations (Oreg 2003). Some people are more likely than others to resist or avoid trying out new things. Such inclinations correspond with the overall profile of the typical Laggard. When a new product is introduced, dispositional resistance often manifests itself in late adoption. Indeed, resistance to change has been found to correlate with time of adoption across a variety of products such as cordless phones, VCRs, and software packages (Oreg, Goldenberg, and Frankel 2005). When multigeneration products are involved, however, the ultimate kind of lagging can be exhibited in a certain generation being skipped altogether. As in Aaron's or the CPA's case, skipping some generations can coincide with the early adoption of other generations. Thus whereas the underlying tendency to resist remains the same, the nature of the product involved (that is, single versus multiple generations) determines how resistance will be manifested, whether in lagging or in the leapfrogging that creates the potential for early adoption.

The Economic Value of the Leapfrogging Effect

Now that we've reached a deeper understanding of the lagging and leapfrogging dynamics and understand that even typical resistors may become leaders of the adoption curve, we should no longer ignore their economical value. Let's return to Aaron, who decided to adopt the newest MP3 player. The direct effect of his purchase on company sales derives from the simple fact that one more person has purchased the product. The size of this effect is obviously negligible. Aaron's actions, however, also entail another, much more consequential effect—the *indirect effect*, more commonly known as the word of mouth that Aaron spreads. Aaron shows his new "toy" to his friends. People see him jogging with his cool new player and may ask him about it. The indirect effect has been shown, in many studies, to be much more considerable than the direct effect, because it initiates something similar to a snowball effect. The more people adopt the innovation, the more powerful the effect (e.g., Hogan, Lemon, and Libai 2003).

When Aaron used his Walkman, his word of mouth about the product was virtually nonexistent, because all around people had already owned newer devices. Contrarily, when Aaron showed up with his new player, his word of mouth now counted as that of an Innovator. The effect of Aaron's opinion is, in fact, probably larger than that of a more typical Innovator because Aaron is a solid, responsible Laggard. He adopts only when he is good and ready, and is little influenced by those around him. Thus Aaron's adoption signals that anyone (even a resistor such as himself) can purchase and easily use the new product. Without realizing it, Aaron becomes a leader.

If people like Aaron, still holding on to their Walkmans, decided to leapfrog today, the newest player would be a likely candidate for their next purchase. Were they to hold on to their Walkmans a little longer, however, they might as well skip even this product and end up purchasing something even newer. For each product, a resistant population exists that will potentially skip, and thus never purchase, that product. This population presents a challenge for firms who want to speed up the adoption process. The challenge is substantial given that these resistors' switching depends, to a great degree, on factors outside of marketers' control, such as when the product breaks down. Yet in many cases, products don't fail all at once, and there is a period of time in which the product could still be used, despite showing signs of failure. During this time period, marketers may have more of an impact on the dispositional resistor.

To begin with, however, how prevalent is the leapfrogging phenomenon? We can speculate about what Aaron would do to replace his old Walkman, but what actually happens to consumers? To provide an initial examination of this phenomenon, we sampled, back in 2006, 105 individuals (66% men and 34% women) between the ages of twenty-six and sixty and asked them about their inventory of portable audio players. Participants were asked whether they owned, and when they had purchased, a portable cassette player, a portable CD player, a MiniDisc, or an MP3 player. Ownership percentages, means, and standard deviations of adoption years are presented in table 5.1. Naturally, an individual can simultaneously own more than one type of player.

The distribution of the leaping distance is presented in figure 5.3. Of those who owned a portable cassette player, 61% moved on to adopt a portable CD player, 6% skipped the CD player and moved on to buy a MiniDisc, 10% leapfrogged over both the CD player and the MiniDisc and bought an MP3 player, and another 23% had not yet purchased any portable player subsequent to the cassette player. Therefore, a total of 16% of those who owned a portable cassette player leapfrogged, with another 23% likely to leap even beyond the MP3 player.

As stated above, almost by definition, the longer individuals hold on to their product, the greater will their leap be. By the time a resistor decides to switch, the most prevalent and accessible product around is likely to be several generations beyond the initial product owned. Accordingly, in our sample, there was a significant correlation ($r = 0.36, p < 0.05$) between

TABLE 5.1 **Portable audio player ownership**

	Portable Cassette Player	Portable CD Player	Mini-Disc	MP3 Player
Percent of participants who have owned product	67	48	14	30
Mean year of adoption (SDs in years are in parentheses)	1995.5 (5.4)	2000.1 (3.9)	2000.9 (1.7)	2003.7 (1.3)

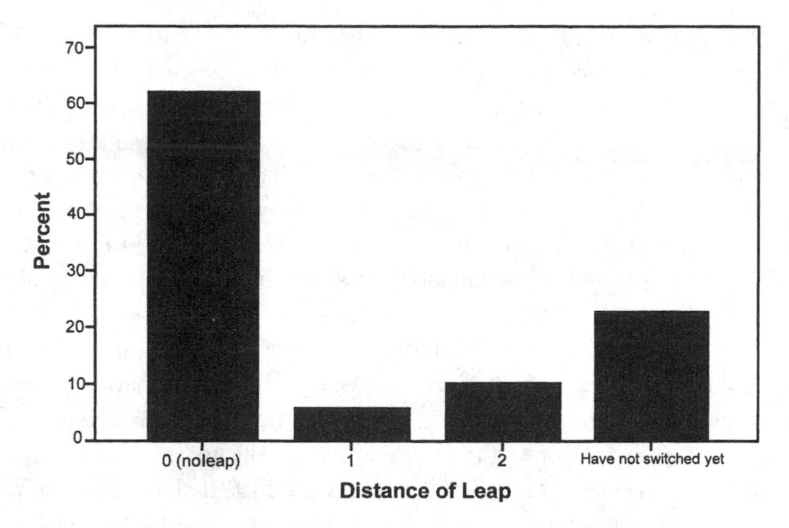

FIGURE 5.3. The leap distance distribution. Figure 1 in Goldenberg and Oreg (2007).

the amount of years participants held on to their portable cassette player and the distance of their leap (the number of products that participants skipped).[2] Whereas the 10% who leaped from the Walkman to the MP3 player have held on to their cassette player for quite some time, and could have likely been chastised by their less resistant friends for not owning a portable CD player, in leaping they progressed beyond the majority of consumers in the market and thus became resistors in an Innovator's disguise. In fact, 22% of those who reported owning an MP3 player in our sample leaped directly from the portable cassette player (that is, skipped the portable CD player and the MiniDisc).

Our purpose in this small study was to gain a preliminary sense of the prevalence of the consumer leapfrogging effect. Considering our sample size, it is likely only a coincidence that the proportion of leapfroggers in our sample precisely corresponds with Rogers's (2003) assessment of the proportion of Laggards in the population, being 16%. Other estimations suggest that Laggards may constitute up to 21% of the market (Mahajan, Muller, and Srivastava 1990). If we accept the contention that lagging and leapfrogging stem from similar sources (for example, resisting the adoption of new products), and that many individuals who are dispositionally disinclined to switch might therefore end up as Innovators on a new product's adoption curve, it is interesting to try to evaluate their economical value to the firm.

The potential gain in addressing these resistors comes from a firm's potential to convince them to leapfrog earlier than they would have naturally. In other words, if the firm owning the latest MP3 players were to somehow convince Aaron, and his like, to abandon their Walkmans earlier, it would benefit from both the direct and indirect effects on its product's sales. Otherwise, the firm would certainly miss out on these benefits at this point in time and likely miss out on them entirely in the case in which these consumers end up adopting a next-generation product, manufactured by another firm.

It is simple to quantify the indirect effect created by resistors' leapfrogging. As noted earlier, given that early-adopting resistors become surrounded by potential adopters, subject to their influence, the indirect impact of an early-adopting resistor is substantially greater (assuming, of course, that the resistor is satisfied with the product) than the direct effect of the actual purchase on the firm's Net Present Value (NPV). Using a firm's diffusion and NPV equations, one can calculate what happens to a new product's adoption curve once a small subgroup of resistors leapfrogs and becomes among the first to adopt that product. Our model employs two parts: in the first (equation 1), we calculate the extent to which a product has been adopted over time, given the market size, the impact of the firm's marketing effort, and the influence of customers' word of mouth. This equation is based on the Bass model (Bass 1969) and allows one to calculate the cumulative product adoption through a simple adoption model where N (t) is the cumulative number of adopters (t is measured in years). The number of annual adopters (usually denoted as n (t), although we do not use this notation here) is a time-dependent number and the derivative of N (t). M represents the market size, p the probability that an individual within the market will adopt in light of the firm's marketing efforts (a.k.a. the external force), and q the probability that an individual will adopt in response to word of mouth (a.k.a. the internal force). In this model, both p and q are assumed to be equal across population and over time.

(1)
$$n(t) = \frac{dN(t)}{dt} = \left(p + q\frac{N(t)}{M}\right) \cdot (M - N(t))$$

In the second part (equation 2), we use the result from the first equation to calculate the expected profit due to the indirect effect created by the word of mouth. We use here a simpler version of the equations that fit the particular context of our study.

(2) $$NPV(t_0,t)=\int_{t_0}^{t} n(s)e^{-d(s-t_0)}ds$$

We are interested in what happens to the product's adoption curve once a small subgroup of resistors leapfrogs and becomes among the first to adopt. For simplicity's sake, let us assume that leapfrogging occurs in the course of the first year following a product's launch, and does not persist in following years. Accordingly, N (t = 1) increases by the number of resistors who leapfrogged. For any given diffusion process, p and q can be derived, and thus n (t) can be calculated. As a next step, n (t) can be substituted into equation 2.

Both calculations in equation (1) and (2) are performed in their discrete form (see Hogan, Lemon, and Libai 2003), in which we use equation (3) for calculating the net present value.

(3) $$\sum_{i=1}^{n} n(t)\frac{k}{(1+d)^{(t_n-t_0)}}$$

K represents a firm's profit from a single adopter (which in our case is fixed to K = 1); d, typically called the *discount rate*, represents the expected annual interest rate a firm is likely to obtain on K (fixed here to be 10%); t is now a discrete variable representing the period number; t_0 represents the time of a product's launch; and n (t) is now a discrete variable, calculated through the discrete form of equation 1, and representing the number of adopters in time t.

We use this model to compare a regular diffusion process with one that includes an additional subgroup of leapfrogging resistors. Let us examine the situation in which 1% of the resistors (assumed to constitute 0.016% of the market, based on the proposed proportion of Laggards in the market), who would have otherwise leapfrogged over the product, were somehow persuaded to expedite their leap and be among the first to adopt.

Figure 5.4 illustrates the acceleration process produced by leapfrogging. A representative diffusion process with p = 0.03, q = 0.3, and M = 90,000 units. As a next step, we assume that 1% of the resistors leapfrog.

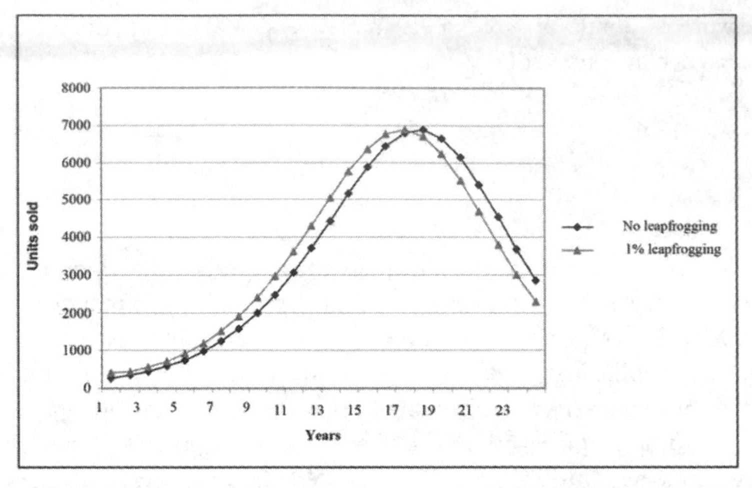

* The gray line denotes a hypothetical curve that reflects product adoption if 1% of Laggards were persuaded to leapfrog. It was assumed that the Laggard population comprises 16% of the market.

FIGURE 5.4. Adoption curves with and without the leapfrogging effect; speeding up the adoption process by leapfrogging. Figure 2 in Goldenberg and Oreg (2007). The gray line denotes a hypothetical curve that reflects product adoption if 1% of Laggards were persuaded to leapfrog. It was assumed that the Laggard population comprises 16% of the market.

As expected, the adoption curve shifts left (the new curve is denoted by a gray line), which indicates that the overall adoption process has accelerated. The NPV of the process increases by 9.2%. Although it is not clear how easy it would be to encourage 1% of the resistors to leapfrog, we believe 1% is nevertheless a conservative estimate.

To quantify this effect further, we examined fifty-four complete adoption processes of clearly innovative products representative of several innovation categories. The coefficients of a diffusion model (Bass) for each process were obtained from Lilien, Rangaswamy, and Van Den Bulte (2000, 300). Among the innovations were agricultural innovations (for example, tractors), medical devices (for example, ultrasound imaging devices), production technology innovations (for example, oxygen steel furnaces), electrical appliances (for example, air conditioners), and consumer electronics items (for example, camcorders).

For each process, we computed two "what if" scenarios in which either 1% or 10% of resistors decided to leapfrog to the product at hand. Because each of the innovations was introduced at a different time, and because not all data sets covered the diffusion process to its conclusion, we used the diffusion coefficients (p, q, and M) provided by Lilien and colleagues to estimate the entire process for the years covered by the data. We assumed that resistors constitute 16% of the population. The NPV calculations, with and without leapfrogging, showed that in the case in which 1% (of resistors) leapfrogged, firms would have increased their profits by 14% on average. In the case in which 10% of resistors leapfrogged, the profits would have increased by up to 89% on average.

To further explore the economic ramifications of the leapfrogging effect, we performed a linear regression analysis with the NPV ratio ([*NPV with leapfrogging – NPV without*] / *NPV without*) as the dependent variable, and the external and internal diffusion coefficients (calculated through the Bass model denoted by p and q), the market potential (M), and a dummy variable representing the two proportions of accelerated leapfroggers (1% or 10%) as independent variables.

Because nonlinear effects were expected, the interactions between p and q with leapfrogging proportions were added (see Aiken and West 1996, for the procedure). In addition, square terms of p and q were added as independent variables to allow for the examination of curvilinear effects. The results of the analysis are presented in table 5.2. Coefficients are standardized, and the adjusted R^2 was 0.41.

Not surprisingly, market potential has no effect on the NPV ratio resulting from leapfrogging (it is evident that market potential does not determine the speed of growth in products' diffusion). It is interesting to note, however, the strong nonlinear effects of the diffusion coefficient,

TABLE 5.2 **Regression analysis predicting NPV**

Parameter	Standardized coefficient	Significance
Leapfrog size	0.23	0.00
P	–0.65	0.00
Q	–0.11	0.30
p^2	0.43	0.00
q^2	0.45	0.00
Leapfrog size x p	–0.20	0.00
Leapfrog size x q	0.21	0.00
Market size (M)	–0.07	0.928

Regression analysis: dependent variable is NPV ratio; all coefficients are standardized; p, q and leapfrogging proportions are centered. The adjusted R^2 is 0.41.

as can be learned from the different coefficient signs. As table 5.2 shows, the NPV ratio decreases with low levels of the external force (p). In other words, in slow process regimes (involving a long left tail before takeoff), the leapfrogging effect is more dramatic, perhaps because it becomes an alternative to marketing forces (p) and enables the process to commence even with a less effective external force. With higher p's, marketing efforts are stronger and compete with the leapfrogging effect, thus attenuating its influence. As the process accelerates, creating a shorter tail before takeoff, the main driver of adoption speed becomes the internal force. In this case, the indirect effect of leapfrogging increases. Contrary to the external force's effect, the internal force's effect (q) is monotonic, as can be seen by the nonsignificant linear term of q, along with the significant squared term (when the squared term is removed from the equation, the linear term becomes significant).

It is not surprising that an increase in the leapfroggers' population size also increases the NPV. However, the results indicate two interaction effects, equivalent in magnitude to the main effects. Analyzing the interactions using median splits (of p and q) corroborated the existence of the nonlinear effects. For low values of p, the effect of leapfrogging is large because it competes with the external force in activating the growth process. As the process accelerates (large p's), the leapfrogging effect decreases, with little difference between cases of high and low levels of leapfrogging (even when the size of the leapfrogging population is multiplied by 10). The significant interaction with q is also consistent with this explanation: the influence of the leapfrogging population size increases with q because of the indirect effect.

This means that the leapfrogging effect is stronger when growth consists of word of mouth, and that if a firm identifies leapfrogging and adoption through word of mouth, fewer marketing expenditures, if any, may be needed. This may also suggest, perhaps counterintuitively, that when an innovation is slow to take off, instead of targeting typical Innovators or main-market consumers, it may be particularly beneficial to shift at least part of the marketing efforts to older-generation consumers and get them to leapfrog.

These findings demonstrate the profound impact on a firm's revenues that leapfrogging may have. Even if a small portion of Laggards can be persuaded to leapfrog earlier than they would have spontaneously, a firm's profits should be expected to increase substantially because of the acceleration in the entire adoption process.

Summary and Conclusions

Whereas the management and marketing literature generally disregards the resistant segment of the consumer population, as do most firms, our findings advocate an entirely different approach to this sector, in particular when product categories involve an inherent process in which new generations substitute old ones. For such products, the Laggard concept needs to be considered more carefully and not be confused with the related yet distinct concept of resistor. We have to define these terms more clearly and explicitly to decipher the real diffusion process of multiple-generation products. In their responses to change or in their patterns of consumption, Laggards and resistors are most worthy of attention and marketing efforts. Whether it is Aaron contemplating what device to use on his morning jogs, the CPA responding to a crisis at his office, or any other conservative consumer, resistors wear many faces and must be approached in various ways to get them to adopt new technologies. Lagging and innovating, trailing and leading, resistors present a complex and rich challenge to firms, who would be remiss to ignore them. With their patterns of adopting and their decision-making processes, resistors highlight the prevalence of resistance in most people's consumption choices. This widespread resistance is no dead end for firms, but rather a rich trove with a code that can be cracked and used, especially in the case of multi-generational products. Convincing even some Laggards to expedite their leaps can bring serious gains to firms and aid in understanding the diverse mechanisms of resistance to innovation. As we discussed in part I of the book, understanding and acknowledging the sources of Laggards' resistance should serve as the starting point to expediting their leaps. For example, as discussed in chapter 1, acknowledging the short-term focus that dispositional resistors possess, marketers should provide sufficient support to help Laggards better cope with the switching costs.

Reorienting marketers' toward Laggards, helping them view Laggards as a resource for increasing the NPV, is a way to turn the disadvantage of resistance to an advantage. Note that our calculations are conservative: we assumed that the influence of a Laggard who adopts a new device is similar to that of an innovator. This is a very conservative assumption because, as we discuss in chapter 7, most consumers are reluctant to consult with Innovators, because of Innovators' high expertise. Consumers often fear that Innovators may not understand the difficulties a technological novice may have. Dispositional Laggards, however, are typically not experts and may therefore be much more approachable. Laggards who adopt early can therefore be very useful to firms.

In the next chapter, we discuss some of the more negative aspects of resistance. Even when a product does not entirely fail, negative word of mouth is a very undesired phenomenon because it slowly decreases the market potential or the market share. We will examine the key mechanisms and dynamics that are associated with the phenomenon of negative word of mouth.

Notes

1. Posted on http://worsethanfailure.com/Comments/Point-of-Fail.aspx, October 18, 2007.

2. For the analysis, the conservative assumption was made that those who have not yet switched products since owning a portable cassette player will switch within the next year. It was also assumed that these late switchers would, on average, leap to the MP3 player. Certainly it is possible that some will switch to earlier generations, yet these are likely to be outnumbered substantially by those who will leap even beyond the MP3.

References

Aiken, Leona S., and Stephan G. West. 1991. *Multiple Regression: Testing and Interpreting Interactions*. New York: Sage.

Bass, Frank. 1969. "A New Product Growth for Model Consumer Durables." *Management Science* 15 (5): 215.

Goldenberg, Jacob, and Shaul Oreg. 2007. "Laggards in Disguise: Resistance to Adopt and the Leapfrogging Effect." *Technological Forecasting and Social Change* 74: 1272–81.

Hogan, John E., Katherine N. Lemon, and Barak Libai. 2003 "What Is the True Value of a Lost Customer?" *Journal of Service Research* 5 (3): 196–208.

Lilien, L. Gary, Arvind Rangaswamy, and Christophe Van Den Bulte. 2000. "Diffusion Models: Managerial Applications and Software." In *New-Product Diffusion Models*, edited by Vijay Mahajan, Eitan Muller, and Yorum Wind. Boston: Kluwer Academic.

Mahajan, V., E. Muller, and R. K. Srivastava. 1990. "Determination of Adopter Categories by Using Innovation Diffusion Models." *Journal of Marketing Research* 27 (2): 37–50.

Oreg, Shaul. 2003. "Resistance to Change: Developing an Individual Differences Measure." *Journal of Applied Psychology* 88 (4): 680–93.

Oreg, Shaul, Jacob Goldenberg, and Rachel Frankel. 2005. "Dispositional Resistance to the Adoption of Innovations." Paper presented at the Annual Meeting of the European Association of Work and Organizational Psychology, Istanbul, Turkey.

Rogers, Everett M. 2003. *Diffusion of Innovations*. 5th ed. New York: Free Press.

Uhl, Kenneth, Roman Andrus, and Lance Poulsen. 1970. "How Are Laggards Different? An Empirical Inquiry." *Journal of Marketing Research* 7: 51–54.

CHAPTER SIX

Resistance and the Dangers of Negative Word of Mouth

Resistance rarely remains within the realms of the individual. It is a theme of communication, which is communicated in a certain manner (Bauer 1995). As it percolates through the market, resistance emerges as a crucial factor in product success or failure. It poses a serious threat for marketers because its transformation from individual to market level is accompanied by a powerful social communication process that is largely invisible and can have a critical effect on new products' fate. This effect results from a virile process known as negative word of mouth (hereafter NWOM), which is another aggregate consequence, or manifestation, of resistance to change (see figure 6.1).

The launch of a new product triggers an interactive spiral of effects in which consumers are exposed to marketing messages and to the positive and negative product-related communications disseminated by other consumers, who may or may not have had any personal experience with the product. NWOM requires little effort by consumers, yet, as studies have shown, has a powerful impact on their attitudes, intentions, and purchase decisions.

In this chapter, we aim to provide a comprehensive picture of the aggregate effects of resistance in which individual-level resistance and market-level resistance are linked by NWOM. This transformational chain of events is triggered by resistance leaders—individuals who resist a product innovation (see part I of this book) and communicate their opinion to others, and by the structure of the social network itself.

The most trivial consequence of the influence of NWOM is naturally a product's failure. When NWOM is significant, an innovation cannot

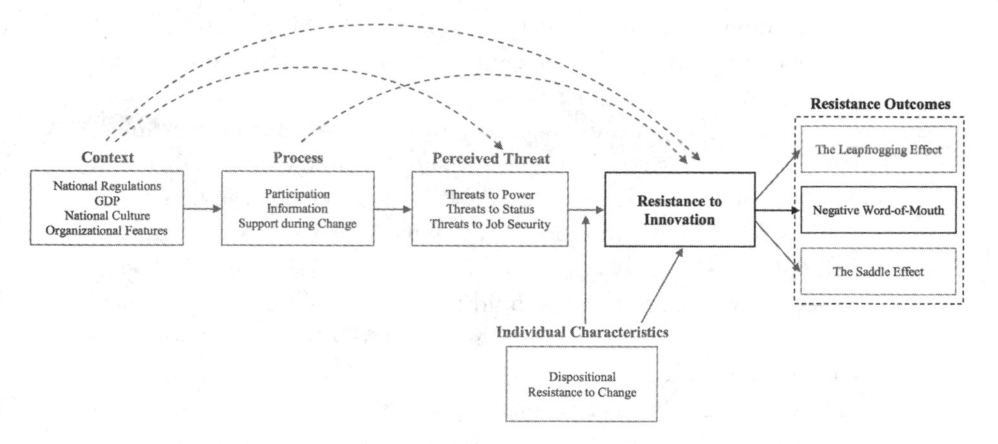

FIGURE 6.1. Resistance and negative word of mouth.

propagate and capture the planned market share. Our focus here, however, is not on failures, but rather on innovations that "suffer" from NWOM but may survive it nonetheless. This is a significant phenomenon, which has attracted little attention in the literature.

Social Communication and Word of Mouth

The role of social communications in determining the fate of a new product has long been recognized. Following diffusion theory (Rogers 1995; Mahajan, Muller, and Wind 2000), new product adoption is viewed as an outcome of persuasion through two forces whose relative importance varies over the product life cycle: (a) the external force, which consists of marketing efforts (such as advertising and sales promotions), and (b) the internal force, comprising interactions among consumers (such as word of mouth, imitation, and externalities). For the sake of simplicity, we refer to all internal elements as "word of mouth" or WOM. Although external forces are crucial for a successful launch and takeoff, their effectiveness is thereafter reduced and internal dynamics become the main force that propels growth (e.g., Rogers 1995; Goldenberg, Libai, and Muller 2001; Day 1970). As more and more people begin to use the product, product-related communications spread through internal force propagation mechanisms outside marketers' control (such as word of mouth and consumer reports). Although adoption models have typically focused on consumer communications that promote new products, one could just as well expect social communications to play an equally critical role in the development of market resistance and opposition.

Resistance is both a cause and an effect. It triggers consumer interactions involving negative information about a product (NWOM), and NWOM creates resistance among others. This is because the flow of information in a market is bilateral, from opinion (or resistance) leaders to their followers, and vice versa (Venkatraman 1989; Yale and Gilly 1995). As noted above, individuals who hold negative attitudes about products and share their opinions with others in the marketplace are known as *resistance leaders*.[1]

Marketers undoubtedly have an intuitive awareness of the critical role of NWOM and the power resistance has in determining products' fates. The business literature contains numerous anecdotal stories about the harm caused by a dissatisfied customer's WOM communications (Hart, Heskett, and Sasser 1990). Consider the following actual case of a consumer electronics company that recently introduced an audio CD protection device. Soon after launch, the company discovered that the product performed poorly in about 2% of the European market. Fixing the problem was not simple, and the firm's executives debated about how much the company should invest to solve the problem. The executives were all aware of the conventional wisdom that "bad news travels fast," but despite the host of sophisticated indexes at their disposal to measure sales and performance, none of them had a good grasp on how to assess the possible effects of the anticipated NWOM on their profits. Some argued that 2% of the market would have negligible economic consequences. Others countered that the dissatisfied customers, who could not be identified in advance, would generate NWOM communications following their poor experience, ultimately causing substantial harm to the firm's profits.

It is not surprising that a reliable measure of NWOM is hard to come by, because the classic diffusion modeling literature had not considered resistance or resistance-based communications. In these classic models, all consumers are potential adopters, even though many will never even consider adopting a given product. Those who eventually adopt the new product are expected to feel certain about their decision, and influence other potential adopters through positive WOM (PWOM). The widespread acceptance of classic adoption models may explain why marketers remain less attuned to resistance and rarely acknowledge or calculate the number of resistors in their potential market. Consequently, marketers may first suffer shock when they encounter the impact of resistance on their product's growth curve as a sharp *thud!* signaling the crash and failure of yet another new product.

The invisibility of NWOM is a second factor that explains marketers' and marketing scholars' relative neglect of resistance and its market-level effects. The social network in which information spreads is a complex system that consists of numerous individual entities, interacting with each other, ultimately generating large-scale, collective, visible, and quantifiable behavior (Anderson 1999; Holland 1995). The positive consumer interactions lead to adoption and growth of sales that are visible at any point in time. Contrarily, NWOM is invisible, leaving no trace in sales curves. As a result, firms have no reliable measure of the NWOM that expresses the number of potential customers who decline product adoption (Haddad-Leibovich, Goldenberg, and Shapira 2011). Like an underwater current, NWOM operates beneath the surface, shrinking the market by transforming potential adopters into nonadopters, emerging only after a mass of nonadopters has become a fait accompli. We describe below some of the findings that have been uncovered about the nature and effects of NWOM.

PWOM and NWOM

Resistance does not necessarily lead to product failure. In fact, resistance is prevalent in most adoption processes. In the course of normal business activity, some customers will be dissatisfied with their products, and some will spread NWOM. As noted above, even in the case of successful products, the invisible process of resistance may cause irreparable damage to market size, which cannot be reversed by follow-up marketing efforts. In other cases, NWOM delays, slows, concentrates, and limits the penetration of new ideas or products and transforms what could have been a critical slowdown in sales into a complete product failure.

Let us further examine the effects of NWOM. Products face dynamics of competition between two countervailing forces (PWOM and NWOM) that are initially ignited by the same trigger—the *external force* (that is, marketing efforts such as advertising). Resistance is sometimes considered a normal consumer response when confronted with innovations. Under such circumstances, adoption begins only after initial consumer resistance is overcome. Most typically, it is the ratio between NWOM and PWOM that determines whether a product will fail and how much money will be lost due to NWOM in those cases where products ultimately succeed.

Sungjoon Nam, Puneet Manchanda, and Pradeep Chintagunta (2010) explored the existence and magnitude of NWOM, as related to the quality of a video-on-demand service. For subscribers, the quality of the service is determined by the number of movies available for viewing at a given time. Nam and colleagues found that contiguous word of mouth affects the adoption behavior of about 8% of subscribers. This effect, however, was found to be asymmetric, with the effect of NWOM being twice as strong as that of PWOM.

Three groups of findings confirm the asymmetric effects of resistance and adoption: (a) negative information is generally more likely to be shared through WOM than positive information, even by nonadopters (Herr, Kardes, and Kim, 1991; Arndt 1967a); (b) more specifically, dissatisfied customers communicate with others more frequently than satisfied ones (Anderson 1998; TARP 1986); and (c) recipients of WOM communications place more weight on negative information (Herr, Kardes, and Kim 1991). The disproportional influence of unfavorable information is also indicated by attribution theory (Mizerski 1982) and, more generally, by the fact that such information is more accessible and diagnostic (Herr, Kardes, and Kim 1991). Other evidence also indicates that negative interactions are more dominant, are associated with higher recall, and have a higher diffusion rate (Herr, Kardes, and Kim 1991; Richins 1983).

Why does negative information have a greater impact on individuals than positive information? Several suggestions have been offered. Michael Kamins, Valerie Folkes, and Lars Perner (1997) proposed that negative information is more vivid than positive information. Furthermore, according to schema incongruity theory, information that is inconsistent with existing schema is processed more diligently (see, for example, Sujan, Bettman, and Sujan 1986). Finally, negative information may be more influential because it is perceived as unbiased and therefore more reliable (Richins 1984; Weinberger, Allen, and Dillon 1981, in DeCarlo, Laczniak, and Ramaswami 1999).

As a result of such asymmetry, the interaction between positive and negative WOM is more complex than a simple rivalry of forces, and counterintuitive results may ensue. David Midgley (1976) recognized such complexity. As we describe below, a detailed model, proposed by Sarit Moldovan and Jacob Goldenberg (2004), illustrates the asymmetric roles of opinion leaders who spread PWOM and resistance leaders who spread NWOM: eliminating opinion leaders (one of several channels through which PWOM is disseminated) will have a merely marginal impact on PWOM communications. Eliminating resistance leaders, on the other hand, will essentially terminate the process of NWOM promulgation.

Identifying the Hidden Enemy

As with the spread of other information, the spread of NWOM is complex and involves a large number of interacting individuals, in a typically indiscernible manner. Although these interactions are ultimately translated into large-scale, collective, visible, and quantifiable behavior (Anderson 1999; Holland 1995), until they are, NWOM is an invisible force, leaving no trace on sales curves (those who don't adopt don't appear on the sales curves). In other words, the transformation of potential adopters into nonadopters is largely invisible to marketers.

As a result, marketers' efforts to measure NWOM have been typically limited to audits of consumer complaints, which constitute a gross underestimate. Not only do dissatisfied customers rarely choose to share their dissatisfaction with the firm, but also NWOM may even be generated by consumers who have never owned the product (Charlett, Garland, and Marr 1995). Attempts at measurement are further confounded by the speed at which negative product-related information can electronically percolate through the market (Ward and Ostrom 2006). With the development of Web 2.0 and various social media sites such as Facebook, LinkedIn, and Twitter, NWOM that might have previously taken weeks to spread can now be transmitted to a great number of individuals almost immediately (Donovan, Mowen, and Chakraborty 1999). Until an accurate method is developed for tracking and measuring the decline in market potential caused by resistance, marketers have little chance of developing effective strategies to counteract its results.

One approach to studying the wide-reaching effects of resistance is to focus on *resistance leaders*—those individuals who spread NWOM. Not all product rejecters (nonadopters) generate NWOM, and not all generators of NWOM are nonadopters. On the contrary, many of those who spread NWOM are dissatisfied product owners. They are thus distinct from "mere" resistors (nonadopters), and, accordingly, the factors that explain their behavior (generation of NWOM) are distinct from, albeit sometimes related to, those that explain nonadoption. Part I of the book was devoted to explaining why individuals resist (do not adopt) innovations. Referencing back to such explanations, we briefly discuss here the research that aimed to characterize resistance leaders.

Although most researchers who study the motivations of NWOM focus on dissatisfied customers, *resistance leaders* are, in fact, a diverse subgroup, comprising individuals who share their negative opinions about products with others. Just as the spread of negative information is similar but not identical in evolution to positive information dissemination, resistance leaders do not necessarily have the qualities typically attributed to *opinion leaders* (that is, those who spread PWOM).

Those who tend to spread NWOM, similar to opinion leaders, have stronger social ties, a higher socioeconomic status, and are more active in formal organizations and social groups in comparison to other consumers (Richins 1987; Singh 1990; Watkins and Liu 1996; Warland, Hermann, and Willits 1975). Several researchers have suggested that resistance leadership is a measurable trait. Marsha Richins (1983), for example, developed a scale of "intent to spread WOM."

Several classes of individual traits have been linked with individuals' motivation to disseminate NWOM, including (a) personality traits such as self-confidence (Lau and Ng 2001) and sociability (Lawther 1978); (b) prosocial attitudes, such as a desire to help others (Arndt 1967b; Dichter 1966; Richins 1984), blame attribution, and attitudes toward complaining (Singh 1990; Lau and Ng 2001); (c) product involvement (Richins and Root-Shaffer 1988), involvement in the purchase decision (Landon 1977; Lau and Ng 2001), problem severity; and (d) situation factors such as proximity of others during dissatisfaction (Bell 1967; Lau and Ng 2001, all cited in Lau and Ng 2001).

There is also evidence for cultural differences in the motivation to initiate NWOM (Lau and Ng 2001). In their study of individual and situational predictors of NWOM, Lau and Ng (2001) found that although several of the factors that affected NWOM behavior were common to individuals from Canada and Singapore, some factors were unique to individuals from a given nationality.

Previous Modeling Efforts—Integrating Individual-Level and Aggregate-Level Diffusion

Identifying and tracking resistance leaders, and resistors in general (see part I of the book), is essential if we are to develop a comprehensive model of innovation—one that integrates both resistance and adoption and individual-level communications with market-level effects. To accurately estimate the effects of resistance on a product's market potential, it is not enough to gain an understanding of how and why individuals resist innovations and communicate their attitudes to others. One also needs to identify and track the silent communications that surreptitiously erode a product's market potential as they develop. To offer marketers what they really need—the tools to predict and counter market resistance—it is essential to integrate both levels of understanding by tracking how resistance plays out and influences the entire market once it emerges.

Since the social structure has a profound effect on information transfer (Barabási 2003), understanding the social system dynamics—social interactions, influence, and interdependence—that drive the growth process is essential in any model aimed at explaining the aggregation of individual NWOM to the network level. It would be difficult to truly understand the role of NWOM in the growth process without taking these dynamics into account. This is, however, a genuine challenge for aggregate modelers because dynamics related to the individual-level communication processes have been generally deemed too complex to model at the aggregate level. As a result, most attempts to conceptualize and understand consumer resistance have focused on either the individual level (for example, psychological models of resistance such as those we describe in chapter 1) or the market level (that is, technology adoption models), but not both.

Recently, several attempts have been made to link individual-level WOM and aggregate-level responses using various agent-based diffusion modeling approaches. The agent-based model used by Sarit Moldovan and Jacob Goldenberg (2004), for example, which aggregates individual responses to estimate market-wide effects of NWOM, underscores how *resistance leaders* can undermine the diffusion of a new product. This model extends the previous conception of NWOM by defining resistance leaders as a subgroup of nonadopters. Corresponding with opinion leaders who disseminate PWOM, resistance leaders are defined as consumers who have a uniquely strong influence on other consumers, and who, after having rejected a product, share their negative product opinions with others in the form of NWOM. Running such a model shows how such kernels of resistance significantly reduce the market of a new product. The more influential these resistance leaders are, the smaller the eventual market.

The model Moldovan and Goldenberg (2004) employed is an extension of the cellular automata model. A general cellular automata model is composed of an array of cells, each having a discrete value (0 and 1, in this example) representing the state of each individual. The transition from

0 to 1 is governed by two probabilistic rules: (1) an external influence that may impact all cells, and (2) interactions among cells. The external influence represents marketing efforts, and the interactions among cells represent internal effects. The model is solved computationally by running a stochastic process in which each individual's probability of adoption (the transition from 0 to 1) is determined, at each period, by the two forces. Although this model is rather simplistic, this oversimplification is somewhat reduced in Moldovan and Goldenberg's model, which extends the cellular automata model in three ways, based on acceptable axioms adopted from diffusion theory (e.g., Mahajan, Muller, and Kerin 1984; Midgley 1976):

1. Consumers (represented by cells) can interact with nonadjacent consumers.
2. Rather than a single homogeneous market, consumers can belong to one of three groups: opinion leaders, resistance leaders, or main-market consumers. The difference between the two types of leaders on the one hand, and the main market on the other, is expressed by the number and intensity of the social ties they maintain: leaders interact with more individuals, and their interactions have a greater influence on others (Rogers 1995; Valente 1995; Valente and Davis 1999; Venkatraman 1999).
3. Consumers (in all groups) may be in one of three states (uninformed, adopter, resistor), rather than the two basic states of adopters and nonadopter. Uninformed consumers do not spread WOM; adopters spread PWOM, and resistors spread NWOM.

In Moldovan and Goldenberg's extended model, all consumers are in an "uninformed" state in the initial stage of the simulation. Consumers become either adopters or resistors after having been affected by internal or external forces. The external force is defined as the advertising message (α), and internal forces may be PWOM/NWOM communicated by a main-market consumer or resistance/opinion leader. Adoption occurs, at a given probability, as a result of PWOM or advertising, whereas rejection occurs as a result of NWOM.

Once exposed to product-related knowledge, main-market consumers may either adopt or reject the innovation (at specified probabilities). Opinion leaders, however, may only adopt the innovation, and resistance leaders may only reject it. This restriction, about the consistency in leaders' behavior, is based on the evidence for the stable, dispositional nature of resistance to change, as we discuss in chapter 1.

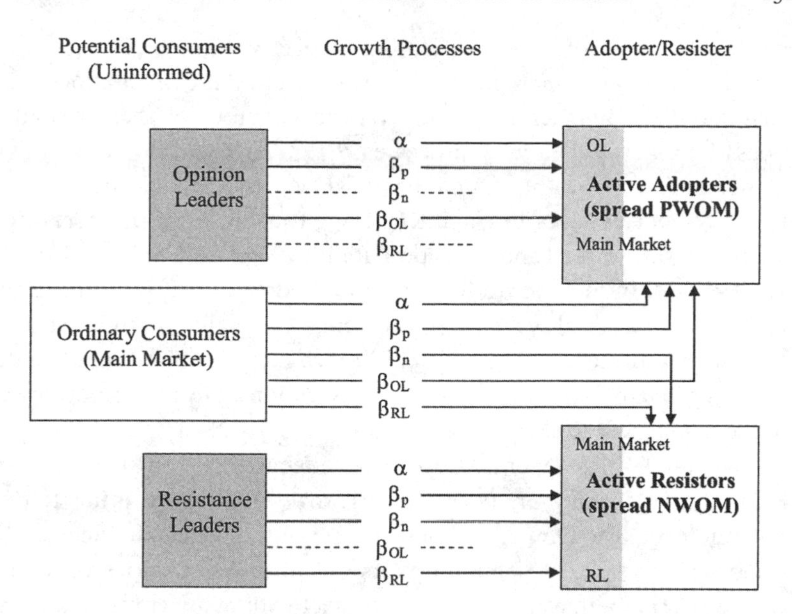

FIGURE 6.2. Model flow diagram. Adapted from figure 1 in Moldovan and Goldenberg (2004). This diagram represents the consumer groups and the grown processes affecting them. Opinion leaders (OL) and resistance leaders (RL) are shaded in gray. Sizes of each group correspond to their relative distribution in the market. Five forces affect the market: advertising (α), and four word-of-mouth effects: positive and negative main-market WOM, opinion leaders' WOM, and resistance leaders' WOM (β_p, β_n, β_{OL}, β_{RL}, respectively). These forces affect each group differently and drive adoption or resistance, consequently disseminating PWOM and NWOM, respectively.

Opinion and resistance leaders are considered two distinct, stable, groups, which, although influencing and influenced by main-market consumers, have no mutual interaction effects. Moreover, because both groups of leaders are small in comparison to the main market, any mutual effects among them would be negligible. This parameter is therefore set to zero.

In the model, resistance leaders are assumed to be activated by either NWOM or by positive information such as advertising or PWOM (this aspect of the model was validated in a separate study). The model flow is based on the following transition rules (the flow diagram is illustrated in figure 6.2).

For each parameter, Modovan and Goldberg (2004) substituted values from ranges consistent with those established in previous studies, performing a total of five manipulations for each parameter, excluding advertising, which was manipulated at seven levels. All combinations were considered in a full factorial design to produce a total of 109,375 simulations of market development processes.

Although research has found that NWOM may have a stronger effect than PWOM (Herr, Kardes, and Kim 1991; Richins 1983), a conservative approach was adopted and equal probabilities were used for both PWOM and NWOM. In the first study, an agent-based model representing 2,500 individuals in a social system is solved computationally through a simulation (see Casti 1999; Wolfram 1984) in which the procedural rules are defined as probabilities for each event. A regression analysis was performed to predict market size (see table 6.1).

The simulations generated two main effects: First, *resistance leaders reduced sales significantly* (see table 6.1, rows 7 and 9), as a function of both their relative number and the strength of their social influence. The compensating force of opinion leaders was insufficient to overcome this influence. The better linked (having contact with many consumers), and the more influential these resistance leaders are, the smaller the eventual market. Note that on average, 20% of the market rejected the innovation, whereas 76% adopted it (another 4% remained uninformed). Disproving popular wisdom, when resistance leaders are present, opinion leaders have the least influence on innovation adoption and ultimate market size. The influence of resistance leaders was found to be four times greater than that of opinion leaders. Market size was hardly affected by increases in the force of opinion leaders' influence or their percentage of the total population (table 6.1, rows 6 and 8). The reason lies in the asymmetry of the process (even with equal *P/N* probabilities). Because PWOM relies on several channels of dissemination (external forces such as advertising, PWOM by main-market consumers, and PWOM by opinion leaders), eliminating one channel—that of opinion leaders—has only a marginal effect on PWOM. Eliminating resistance leaders, on the other hand, essentially terminates the NWOM promulgation process. The combination of both opinion leaders and resistance leaders in a simulation creates a joint effect—the market shrinks by 14% compared to its maximum potential.

Second, *advertising had a limited impact on the ultimate market size* (table 6.1, rows 1 and 2) and, at high levels, even reduced market size. Intuitively, more advertising attracts more consumers to consider adopting the innovation. Advertising, however, also activates the market's resistance

TABLE 6.1 **Study 1—Impact on Market Size (Moldovan and Goldenberg, 2004)**

	Market size	t
Advertising effect	0.14	17.68
Advertising effect ^2	−0.05	−6.32
Ordinary PWOM effect	1.82	186.55
Ordinary PWOM effect ^2	−1.25	−128.06
Ordinary NWOM effect	−0.29	−157.53
OL PWOM effect	0.07	35.38
RL NWOM effect	−0.26	−137.69
Number of OL	0.04	23.88
Number of RL	−0.19	−104.21
Adjusted R^2	0.62	

Based on 109,375 simulations. All variables are significant at $p < 0.001$ level. OLS parameters are standardized. The dependent variable is the size of the market at the end of the simulation. OL and RL denote opinion leaders and resistance leaders, respectively.

leaders, who (like opinion leaders) are highly attentive to advertising, are well connected (Rogers 1995; Venkatraman 1989), and respond by increasing the effect of the negative internal force. Main-market consumers' PWOM was the most influential factor affecting market size. This effect was much stronger than that of NWOM (table 6.1, rows 3 and 5). The effect of PWOM, however, is nonlinear (table 6.1, rows 3 and 4) and, similar to advertising, activates resistance leaders at high levels.

Moldovan and Goldenberg (2004) designed a second study to explore how one may reduce the detrimental effect resistance leaders have on market growth and size. A series of scenarios was developed using sets of parameters that correspond with a variety of marketing strategies. Each scenario creates a "would-be world" (see Casti 1999) through which the outcomes of each strategy can be evaluated. One possible solution to counteract the restricting effect that resistance leaders have on opinion leaders' positive market impact is to grant opinion leaders an advantage. For example, earlier activation of the opinion leaders may enhance their impact before resistance leaders have a chance to erode the market.

Three separate analyses of simulation results were performed: first, a comparison of the final market size in the two designs; second, a regression analysis using the two designs for a total of 218,750 simulations, including a dummy variable reflecting the existence of preactivated opinion leaders; and third, a comparison of the various effects across the two studies.

Results showed that market size increased when opinion leaders were activated prior to the initiation of the advertising efforts. In fact, preactivation of opinion leaders moderated the overall impact of the internal forces. PWOM and NWOM of ordinary consumers, with no preactivation, increased the market by 31% and diminished it by 18%, respectively, whereas in the preactivation condition, their influence was 19% and 8%, respectively. Opinion leaders increased the market by 10% (4% in the non-pre-activation condition), whereas resistance leaders reduced market size by 10% (16% in the non-pre-activation condition).

In summary, Moldovan and Goldenberg's (2004) studies highlight the detrimental effect that kernels of resistance have on the diffusion of innovations. Even in the case of successful products, resistance, though invisible, may cause irreversible damage to market size. This damage cannot be rectified by opinion leaders or by an increase in marketing efforts. The latter has no more than a limited effect, as it concurrently activates both opinion and resistance leaders. To mitigate the negative effects of NWOM on market growth, opinion leaders need to be brought into play in advance of unfocused marketing efforts.

More recently, Goldenberg, Libai, Moldovan, and Muller (2007) developed the "Bad News NPV" model, which they use as a means to understand the economic implications of NWOM. Their model relies on a diffusion-based approach in which WOM is considered an interpersonal interaction, designed to reduce consumers' uncertainty. As such, it is rather unique in its use of an extended small-world model (Watts 1999) for exploring how NWOM evolves at the individual level and disseminates over social networks to affect aggregate sales.

The Bad News NPV model is a dynamic generalization of the small-world approach that takes into account the existence of various levels of communication. It thus simulates a market in which information spreads when consumers interact with each other, using both strong ties within their own social system and weak ties with other networks (Granovetter 1973). The simulated market is based on several assumptions. First, the strong ties within the network are distinct from the weak ties outside the network. A single strong-tie connection conveys more reliable information and therefore has a greater chance of influencing potential adopters. Second, whereas the strong-tie structure inside each social system is fixed, the weak-tie structure is dynamic. In each period, weak ties are randomly reassigned, so that the new structure of the weak-tie network differs from that of the previous period. The latter assumption reflects the dynamic character of weak ties as described in the original work of Granovetter

(1973) and in subsequent literature. The uniqueness of weak ties lies in their variability from one period to another.

Whereas Granovetter's (1973) seminal work incited a series of studies on the strength of weak ties that connect various networks and catalyze information assimilation (e.g., Cross and Levin 2004; Rindfleisch and Moorman 2001; Brown and Reingen 1987), these analyses disregard the existence of NWOM. Bell and Song (2005) later introduced a utility-based approach, with the interaction appearing as thresholds, whereas Shaikh, Rangaswamy, and Balakrishnan (2005) used yet another approach. Their classic small-world model begins with close neighbors fully connected by strong links, with each link being replaced with a weak link to another individual, selected randomly. Thus, they can control the connectedness of their network by rewiring the probability that determines the quantity of weak and strong ties. The Bad News NPV study was the first to explore the strength of weak ties under the assumption that NWOM exists.

In the Bad News NVP model's diffusion process, (a) individuals in the market are led to adopt the new product as a result of external and internal forces. The diffusion parameters p and q (the impact on adoption of marketing efforts and communications among consumers, respectively) reflect diffusion processes of the type actually witnessed and recorded in markets. (b) Individuals can disseminate PWOM or NWOM. (c) NWOM is spread both by individuals who had experience with the product (dissatisfied adopters) and those who did not (resistors). (d) Individuals are exposed to positive information, negative information, both, or neither. (e) Individuals are connected to each other by either weak or strong ties.

To learn about the roles of network and social structure, and firm actions, a linear regression was performed with the dependent variable being the NPV ratio. In table 6.2, we present the results of the OLS regression and results of the same regression with the addition of a squared parameter for testing the possibility of a nonlinear advertising effect.

Clearly, the percentage of dissatisfied buyers (d) has the strongest effect on the NPV ratio, as measured by the standardized coefficient. For every percentage point of disappointed customers, the loss due exclusively to NWOM increases by 1.82%. Had a repeat-purchase model been used, in which these customers would also cease to repurchase, the effect would have been even stronger.

Another interesting result of this analysis concerns the strength of weak ties. The coefficients of both interpersonal parameters are negative, indicating that the higher the parameters, the stronger the effect of

TABLE 6.2 **Regression results when the dependent variable is the NPV ratio**

Parameter	Coefficient	Standardized coefficient	Coefficient	Standardized coefficient
	Linear advertising		Concave advertising	
r (ratio of strong ties to weak ties)	0.012	0.055	0.012	0.055
d (% disappointed adopters)	−1.822	−0.772	−1.822	−0.772
p (advertising effect)	17.619	0.336	31.943	0.609
q_s (strong-tie WOM)	−0.7	−0.089	−0.7	−0.089
q_w (weak-tie WOM)	−6.57	−0.139	−6.57	−0.139
p squared			−1,302.17	−0.281
Adjusted R^2	73.9%		74.3%	

All coefficients significant at $p < 0.001$.

NWOM on profits. This result may be surprising given the popular view that weak ties would have only a weak effect on the spread of information (Granovetter 1973; Brown and Reingen 1987). We offer two reasons for this result: First, offering a microexplanation, weak ties may become more relevant once people exhaust their strong-tie potential, after which they will share their knowledge with their entire personal network. Second, a macroexplanation is that in the case of a large number of small networks, weak ties are responsible for the activation of networks, and their effect competes with the effect of advertising (Goldenberg, Libai, and Muller 2001). Moreover, weak ties seem to have an especially important role in the presence of NWOM, in which both strong and weak ties disseminate PWOM and NWOM. This difference implies that, given their capacity to erode firms' profits in the presence of NWOM, weak ties should indeed not be underestimated.

Another manifestation of the ambiguous power of weak ties is in the positive effect of the ratio of strong ties to weak ties on the NPV ratio. This implies that when keeping the total number of ties constant, increases in the number of strong ties dissipate the destructive effects of NWOM.

When the level of weak-tie WOM communications is high, their ability to spread NWOM is strong enough so as to reduce the NPV ratio, regardless of the effect of strong-tie WOM. When the weak-tie effect is weaker, the NPV ratio is higher and NWOM has a smaller effect on the market. In this situation, a combination of low weak-tie WOM and low strong-tie WOM creates a high NPV, because WOM is hardly disseminated, and information is therefore communicated mainly through the positive effect of advertising. As strong-tie WOM communications increase, the NPV ratio decreases, due to the propagation of PWOM (although PWOM is also disseminated, it has a less visible effect, because advertising can have an effect that is comparable to that of PWOM). At a very high level of strong-tie WOM (and low weak-tie WOM), the NPV ratio increases again. In this situation, positive strong-tie WOM starts to show its effect on the NPV by accelerating the diffusion process, overcoming the effect of NWOM.

Furthermore, the Bad News NPV study distinguished between successful and failed products and found that the percentage of disappointed customers is what most strongly instigates failure. The strongest positive parameter is the advertising coefficient *p*, which increases chances of success. Interestingly, strong- and weak-tie WOM coefficients have opposite effects. Weak ties increase the chances that a product will fail, as they are responsible for activating negative networks, whereas strong ties still support a product's growth (although it is less likely to become a smashing success).

It was also shown that advertising plays a deceptive role. Used excessively, it yields a promising effect, at least initially. Marketers are wrong to assume, however, that this effect will necessarily persist (Mahajan, Muller, and Kerin 1984). Not only does the advertising effect dissipate over time; in the presence of NWOM, it actually plays a destructive role in new product growth.

Adding yet another dimension to their exploration, Goldenberg, Libai, Moldovan, and Muller (2007) explored the precise mechanism of market destruction. Specifically, they examined how NWOM actually destroys growth. Because the spread of NWOM in real life is mostly invisible, it is easier to trace its effects in a complex model. Two factors reflect the spread of NWOM. The first is the number of product rejecters in the market, and the second is the number of unactivated strong-tie networks (in which no member has adopted the product). These strong-tie networks not only stall the adoption process, but even worse, if they are exposed to NWOM before a single network member adopts, NWOM also might infect the entire network and effectively block adoption by all network members.

The number of rejecters and the number of unactivated networks were found to be the two main mechanisms of market destruction, beyond the percentage of disappointed consumers (as explained, by the end of the process, this direct effect also becomes indirect). Surprisingly, there were no direct positive effects on the NPV ratio, which compares processes with NWOM to the same processes without NWOM. The only positive effects were through decreasing the number of rejecters (by advertising), or by activating new networks (by advertising or through WOM).

These findings highlight the ambiguous role that weak ties (*qw*) play in a market. When promoting both PWOM and NWOM among networks, weak ties activate new networks, yet they may also block them entirely (see details below). Weak ties therefore have two indirect effects on the NPV: while they increase the number of rejecters, they also activate new networks, thus increasing the NPV.

Contrary to weak ties, strong ties affect only the specific network and have no effect on the infection of new networks with NWOM. Once an unactivated network is affected by NWOM (as a result of weak ties), however, strong ties will infect other members of the network with NWOM and may block adoption altogether. They therefore have a strong effect on network blockage. Strong ties do not have any effect on the number of rejecters, which is an amalgamation of network members' strong PWOM. This, in turn, increases the number of adopters before any rejection occurs. Thus, when focusing on the destructive effect of NWOM represented by the NPV ratio, strong ties seem to mainly block networks with NWOM.

The percentage of disappointed adopters (*d*) has an interesting effect. Disappointed consumers increase the number of rejecters, and, in turn, the number of unactivated nets due to these rejecters. Yet, at the same time, the level of disappointment directly reduces the NPV ratio. As explained above, however, at the end of the process, the direct effect on the NPV ratio disappears. This suggests that the effect of disappointed consumers is not due to their NWOM as much as to the spread of second-hand WOM, which, by the end of the process, increases rejection. Companies may therefore not experience the full impact of their disappointed consumers until the end of the process, by which time the market will have become completely resistant. In addition, once NWOM begins, it may be too late for companies to stop it by improving the product.

The Bad News NVP model was the first to demonstrate the potential role of weak ties in the presence of NWOM, and its findings highlight the destructive power of NWOM, accounting for an increase in the number of rejecters and the probability of a product's failure. This analysis shows that NWOM is more significantly affected by weak rather than strong ties among individuals. Weak ties' ability to link remote networks creates a unique opportunity for the transfer of negative information, which can have a devastating effect on the fate of a new product.

How NWOM Affects Individual Purchase Decisions

Using the concept of NWOM, the models described above offer a glimpse into the transformation of resistance from an individual-level attitude to a market-level phenomenon. These models assume that an "influence process" of some kind occurs when a consumer is exposed to NWOM, probably involving cognitive modification of brand evaluations (Bone 1995; Herr, Kardes, and Kim 1991).

Some researchers (DeCarlo, Laczniak, and Ramaswami 1999, following work by Hilton 1995) suggested integrating two theoretical approaches to explain the effects of NWOM on consumers' brand evaluations: cognitive response theory and attribution theory. The central theme underlying attribution theory is that individuals make sense of their social world by making causal attributions of behaviors and events (Heider 1958; Kelley 1972). In the context of NWOM, recipients try to explain the motivation of the source of the NWOM. These attributed motivations further mediate the impact of NWOM messages on recipients' brand evaluations and purchase intentions (DeCarlo, Laczniak, and Ramaswami 1999). Although the social foundation of the process has been established, a greater understanding is needed of *how* NWOM is received and interpreted by the consumer, and how it influences consumers' product attitudes, purchase intentions, and actual purchase behavior.

Prevention, Mitigation, and Managerial Implications

From a managerial viewpoint, managers would be well advised to acknowledge the destructive power of NWOM, as even a small percentage of dissatisfied consumers could cause considerable damage to long-term profits. These dissatisfied consumers create an invisible network of product decline, which could have a fatal impact on product success. Specifically, managers are cautioned to:

- *Beware of the nonlinear effect of advertising.* Managers might be tempted to conclude that in the presence of NWOM, the firm should increase advertising to combat and perhaps even eradicate NWOM. The optimal level of advertising, however, is highly affected by the process of WOM: while increasing the number of adopters, advertising also indirectly increases the number of disappointed customers. This yields an earlier start—at least in part—of the NWOM process. Too much advertising may therefore launch an initial wave of NWOM that, with time, could gather greater momentum (due to its logarithmic growth) than that acquired with decreased advertising efforts. Consequently, too much advertising may, paradoxically, negatively affect profitability. Similarly, the presence of weak ties, which are beneficial to the firm under normal circumstances, might adversely affect it in the presence of dissatisfied consumers.

- *Be proactive in mitigating the effects of NWOM.* Since recipients' response to WOM is influenced by impressions of the target object held in memory (Herr, Kardes, and Kim 1991; Laczniak, DeCarlo, and Ramaswami 2001), the impact of WOM messages may be moderated by brand-name strength. DeCarlo and colleagues (2007) examined the effect of two components of brand equity—familiarity and image—and concluded that the effect of NWOM will be weaker as a store's image is more positive. Following Levy and Weitz (2004), they therefore advise bolstering retailer image and reinforcing familiarity to help protect a store from the potentially harmful effects of negative information.
- *Identify and use opinion leaders.* As demonstrated in one of the studies by Moldovan and Goldenberg (2004), when opinion leaders are activated in advance of unfocused marketing messages (such as advertising), market size increases significantly. Under this condition, opinion leaders' PWOM has a stronger effect, whereas the impact of WOM by resistance leaders and main-market consumers has weaker effects.

  Identifying opinion leaders is not, however, a simple task. Although opinion leaders may have several discriminating attributes, these may differ by innovation domain (Rogers 1995; Flynn, Goldsmith, and Eastman 1996; Venkatraman 1989). It is even more difficult to prevent the emergence of opposition to new innovations. As a start, part I of the book is devoted to unraveling the factors that explain why such opposition exists in the first place. For example, as we suggest in chapter 2, resistance may occur when an innovation deviates from accepted social norms, which threatens opinion leaders' status, or when an innovation requires the acquisition of new knowledge, which threatens opinion leaders' expertise (Leonard-Barton 1985; Rogers 1995; Mukherjee and Hoyer 2001).
- *Identify potential resistance leaders.* Companies should be more closely attuned to consumers who are likely to engage in NWOM. Research on resistance leaders may provide sufficient grounds for at least a broad identification of this group. For example, resistance leaders are high in self-confidence and sociability (Lau and Ng 2001). Furthermore, as we discussed in chapter 1, resistance

leaders may be dispositionally resistant to change. Like opinion leaders, however, substantial variance is likely to exist among resistance leaders, who are likely to vary by product category: companies that sell products that require a high level of product and purchase involvement should be aware of the heightened risk of NWOM.

- *Listen to consumers.* Companies should make it easy for consumers to complain directly and be responsive to consumers and their dissatisfaction. Consumers engage in NWOM when they feel that there is nothing to be gained by complaining (Lau and Ng 2001). Furthermore, because NWOM is affected by others' proximity at the time of the dissatisfying experience, companies can train employees to obtain feedback at the point of consumption to resolve and dispel dissatisfaction as soon as possible.
- *Finally, resistance is not necessarily a negative phenomenon.* Instead of conceptualizing resistance as an irrational consumer attribute, it would be more effective to view it as an opportunity to improve products and their dissemination. We suggest that NWOM, if identified, can direct manufacturers' attention to important aspects of the innovation that require immediate resolution; resistance is an early warning system about issues and expectations that require attention. Resistance constitutes an evaluation of the innovation process at a relatively early stage, and it motivates changes to new products.

A final word of caution: resistance leaders are not the sole source of negative effects on new products' sales curves. Purchase decisions may be influenced by the negative effects of imitation, social learning, and social pressure, among others (see Van den Bulte and Stremersch 2004). One example of negative contagion occurs when consumers are adversely affected by the mere adoption of a new product by other consumers whose adoption reduces the social utility of an earlier product. Joshi, Reibstein, and Zhang (2006) illustrated this negative effect of the adopters of the Porsche SUV (Cayenne) on the potential adopters of traditional Porsche roadsters. Based on a meta-analysis of aggregate diffusion models, Van den Bulte and Stremersch (2004) suggest that imitation effects may overall be stronger than WOM effects in the growth of markets for new products. In modeling NWOM, we are therefore underestimating the impact of resistance on the market.

The effects of resistance illustrate the power of the individual in the marketplace. Although many subscribe to the view that the world is run by large corporations, in which consumers are cynically manipulated by the media and advertising, consumers have a greater impact than they realize on products' overall market potential, firms' profits, and even the rise and fall of new products. For precisely this reason, the power of negative word of mouth is a meaningful force that should not be overlooked by marketers.

In the next chapter, we discuss a different situation in which resistance does not necessarily lead to negative word of mouth but rather to a slower adoption rate in a (large) submarket. This, in turn, may lead to a dangerous yet fascinating phenomenon of an atypical product life cycle characterized by a double-peaked curve.

Note

1. Note that adopters who spread NWOM are not resistance leaders. They are usually dissatisfied consumers.

References

Anderson, E. 1998. "Customer Satisfaction and Word-of-Mouth." *Journal of Service Research* 1 (1): 5–17.

Anderson, P. 1999. "Complexity Theory and Organization Science." *Organization Science* 10 (3): 216–32.

Arndt, Johan. 1967a. "Role of Product-Related Conversations in the Diffusion of a New Product." *Journal of Marketing Research* 4 (August): 291–95.

———. 1967b. "Word of Mouth Advertising and Informal Communication." In *Risk Taking and Information Handling in Consumer Behavior*, edited by D. F. Cox, 188–239. Boston: Division of Research, Graduate School of Business Administration, Harvard University.

Barabási, Albert-László. 2003. *Linked: How Everything Is Connected to Everything Else and What It Means for Business, Science, and Everyday Life*. New York: Plume Penguin Books.

Bauer, M. 1995. "Resistance to New Technology and Its Effects on Nuclear Power, Information Technology, and Biotechnology." In *Resistance to New Technology: Nuclear Power, Information Technology and Biotechnology*, edited by M. Bauer, 1–41. Cambridge: Cambridge University Press.

Bell, D. R., and S. Song. 2005. *Neighborhood Effects and Trial on the Internet: Evidence from Online Grocery Retailing*. Working paper, the Wharton School.

Bell, G. D. 1967. "Self-Confidence, Persuasibility, and Cognitive Dissonance among Automobile Buyers." *Journal of Marketing Research* 4: 46–52.

Bone, P. F. 1995. "Word-of-Mouth Effects on Short-Term and Long-Term Product Judgements." *Journal of Business Research* 69 (1): 213–23.

Brown, Jacqueline J., and Peter H. Reingen. 1987. "Social Ties and Word-of-Mouth Referral Behavior." *Journal of Consumer Research* 14 (December): 350–62.

Casti, J. 1999. "Firm Forecast." *New Scientist* 162 (April): 42–46.

Charlett, D., R. Garland, and N. Marr. 1995. "How Damaging Is Negative Word-of-Mouth?" *Marketing Bulletin* 6: 42–50.

Cross, R., and D. Z. Levin. 2004. "The Strength of Weak Ties You Can Trust: The Mediating Role of Trust in Effective Knowledge Transfer." *Management Science* 50 (11): 1477–90.

Day, George S. 1970. "Using Attitude Change Measures to Evaluate New Product Introductions." *Journal of Marketing Research* 7: 474–82.

DeCarlo, Thomas E., Russell N. Laczniak, Carol M. Motley, and Sridhar N. Ramaswami. 2007. "Influence of Image and Familiarity on Consumer Response to Negative Word-of-Mouth Communication about Retail Entities." *Journal of Marketing Theory and Practice* 15 (1): 41–51.

DeCarlo, Thomas E., Russell N. Laczniak, and Sridhar N. Ramaswami. 1999. "Toward an Understanding of Consumers' Processing of Negative Word-of-Mouth Communication: An Integrated Model." American Marketing Association, Conference Proceedings 10: 159–60.

Dent, Eric B., and Susan Galloway Goldberg. 1999. "Challenging 'resistance to change.'" *Journal of Applied Behavioral Science* 35 (1): 25–41.

Dichter, Ernest. 1966. "How Word-of-Mouth Advertising Works." *Harvard Business Review* 44 (6): 147–66.

Donovan, T., J. C. Mowen, and G. Chakraborty. 1999. "Urban Legends: The Word-of-Mouth Communication of Corporate Morality." *Marketing Letters* 10 (1): 23–34.

Flynn, Leisa R., Ronald E. Goldsmith, and Jacqueline K. Eastman. 1996. "Opinion Leaders and Opinion Seekers: Two New Measurement Scales." *Academy of Marketing Science* 24 (2): 137–47.

Goldenberg, Jacob, Barak Libai, Sarit Moldovan, and Eitan Muller. 2007. "The NPV of Bad News." *International Journal of Research in Marketing* 24: 186–200.

Goldenberg, Jacob, Barak Libai, and Eitan Muller. 2001. "Using Complex Systems Analysis to Advance Marketing Theory Development: Modeling Heterogeneity Effects on New Product Growth through Stochastic Cellular Automata." *Academy of Marketing Science Review* 9.

Granovetter, Mark S. 1973. "The Strength of Weak Ties." *American Journal of Sociology* 78 (May): 1360–80.

Haddad-Leibovich, Keren, Jacob Goldenberg, and Daniel Shapira. 2011. Estimating Decliners' Growth. Manuscript under review.

Hart, C., J. L. Heskett, and E. W. Sasser. 1990. "The Profitable Art of Service Recovery." *Harvard Business Review* 68 (4): 148–56.

Heider, Fritz. 1958. *The Psychology of Interpersonal Relations*. New York: Wiley.

Herr, Paul M., Frank R. Kardes, and John Kim. 1991. "Effects of Word-of-Mouth and Product-Attribute Information on Persuasion: An Accessibility-Diagnosticity Perspective." *Journal of Consumer Research* 17 (4): 454–62.

Hilton, Denis J. 1995. "The Social Context of Reasoning: Conversational Inference and Rational Judgment." *Psychological Bulletin* 118 (2): 248–71.

Holland, J. H. 1995. *Hidden Order*. New York: Helix Books.

Joshi, Y., D. Reibstein, and J. Z. Zhang. 2006. *Optimal Entry Timing in Markets with Social Influence*. Working Paper, the Wharton School.

Kamins, Michael A., Valerie S. Folkes, and Lars Perner. 1997. "Consumer Responses to Rumors: Good News, Bad News." *Journal of Consumer Psychology* 6: 165–87.

Kelley, Harold H. 1972. "Causal Schemata and the Attribution Process." In *Attribution: Perceiving the Causes of Behavior*, edited by E. E. Hones, D. E. Kanouse, H. H. Kelley, R. E. Nisbett, S. Valins, and B. Weiner, 151–74 Morristown, NJ: General Learning Press.

Laczniak, Russell N., Thomas E. DeCarlo, and Sridhar N. Ramaswami. 2001. "Consumers' Responses to Negative Word-of-Mouth Communication: An Attribution Theory Perspective." *Journal of Consumer Psychology* 11 (1): 57–73.

Landon, E. L. 1977. "A Model of Consumer Complaint Behavior." In *Consumer Satisfaction, Dissatisfaction, and Complaining Behavior*, edited by A. R. L. Day, 31–35. Bloomington: Indiana University.

Lau, G. T., and S. Ng. 2001. "Individual and Situational Factors Influencing Negative Word-of-Mouth Behaviour." *Canadian Journal of Administrative Sciences* 18 (3): 163–78.

Lawther, K. 1978. "Social Integration of the Elderly Consumer: Unfairness, Complaint Actions, and Information Usage." In *Proceedings of the* 1978 *Educators' Conference*, edited by S. C. Jain, 341–45. Chicago: American Marketing Association.

Leonard-Barton, Dorothy. 1985. "Experts as Negative Opinion Leaders in the Diffusion of a Technological Innovation." *Journal of Consumer Research* 11 (4): 914–26.

Levy, Michael, and Barton Weitz. 2004. *Retailing Management*. New York: McGraw-Hill/Irwin.

Mahajan Vijay, Eitan Muller, and Roger A. Kerin. 1984. "Introduction Strategy for New Products with Positive and Negative Word of Mouth." *Management Science* 30: 1389–404.

Mahajan, Vijay, Eitan Muller, and Yoram Wind. 2000. *New-Product Diffusion Models*. New York: Kluwer Academic.

Midgley, David F. 1976. "A Simple Mathematical Theory of Innovative Behavior." *Journal of Consumer Research* 3.

Mizerski, R. W. 1982. "An Attribution Explanation of the Disproportionate Influence of Unfavorable Information." *Journal of Consumer Research* 9 (3): 301–10.

Moldovan Sarit, and Jacob Goldenberg. 2004. "Cellular Automata Modeling of Resistance to Innovations: Effects and Solutions." *Technological Forecasting and Social Change* 71 (5): 425–42.

Mukherjee, Ashesh, and Wayne D. Hoyer. 2001. "The Effect of Novel Attributes on Product Evaluation." *Journal of Consumer Research* 28 (December): 462–72.

Nam, Sungjoon, Puneet Manchanda, and Pradeep K. Chintagunta. 2010. "The Effect of Signal Quality and Contiguous Word of Mouth on Customer Acquisition for a Video-on-Demand Service." *Marketing Science* 29 (4): 690–700.

Nord, Walter R., and John M. Jermier. 1994. "Overcoming Resistance to Resistance: Insights from a Study of the Shadows." *Public Administration Quarterly* 17 (4): 396.

Richins, Marsha L. 1983. "Negative Word-of-Mouth by Dissatisfied Consumers: A Pilot Study." *Journal of Marketing* 47 (Winter): 68–78.

———. 1984. "Word of Mouth Communication as Negative Information." In *Advances in Consumer Research*, vol. 11, edited by Thomas Kinnear. Provo, UT: Association for Consumer Research, 697–702.

———. 1987. "A Multivariate Analysis of Responsiveness to Dissatisfaction." *Journal of the Academy of Marketing Science* 15 (3): 24–31.

Richins, Marsha L., and Teri Root-Shaffer. 1988. "The Role of Involvement and Opinion Leadership in Consumer Word-of-Mouth: An Implicit Model Made Explicit." *Advances in Consumer Research* 15: 32–36.

Rindfleisch, Aric, and Christine Moorman. 2001. "The Acquisition and Utilization of Information in New Product Alliances: A Strength-of-Ties Perspective." *Journal of Marketing* 65 (2): 1–18.

Rogers, E. M. 1995. *The Diffusion of Innovations*. 4th ed. New York: Free Press.

Shaikh, Nazrul I., Arvind Rangaswamy, and Anant Balakrishnan. 2005. *Modeling the Diffusion of Innovations Using Small-World Networks*. Working paper, Penn State University.

Singh, J. 1990. "Voice, Exit, and Negative Word-of-Mouth Behaviors: An Investigation across Three Service Categories." *Journal of the Academy of Marketing Science* 18 (1): 1–16.

Sujan, M., J. R. Bettman, and H. Sujan. 1986. "Effects of Consumer Expectations on Information Processing in Selling Encounters." *Journal of Marketing Research* 23: 346–53.

TARP, Technical Assistance Research Programmes. 1986. *Consumer Complaint Handling in America: An Update Study*. Washington, DC: White House Office of Consumer Affairs.

Valente, Thomas W. 1995. *Network Models of the Diffusion of Innovations*. Cresskill, NJ: Hampton Press.

Valente, Thomas W., and R. L. Davis. 1999. "Accelerating the Diffusion of Innovations Using Opinion Leaders." *Annals of the American Academy of Political and Social Science* 566: 55–67.

Van den Bulte, Christophe, and Stefan Stremersch. 2004. "Social Contagion and Income Heterogeneity in New Product Diffusion: A Meta-Analytic Test." *Marketing Science* 23: 530–44.

Venkatraman, Meera P. 1989. "Opinion Leaders, Adopters, and Communicative Adopters: A Role Analysis." *Psychology and Marketing* 6 (1): 51–68.

Ward, J. C., and A. L. Ostrom. 2006. "Complaining to the Masses: The Role of Protest Framing on Customer-Created Complaint Web Sites." *Journal of Consumer Research* 33 (2): 220–30.

Warland, Rex H., Robert O. Hermann, and Jane Willits. 1975. "Dissatisfied Consumers: Who Gets Upset and Who Takes What Action?" *Journal of Consumer Affairs* 9 (Winter): 148–63.

Watkins, H. S., and R. Liu. 1996. "Collectivism, Individualism, and In-Group Membership: Implications for Consumer Complaining Behavior in Multicultural Contexts." *Journal of International Consumer Marketing* 8 (3, 4): 69–96.

Watts, D. J. 1999. *Small Worlds*. Princeton, NJ: Princeton University Press.

Weinberger, Marc G., Chris T. Allen, and William R. Dillon. 1981. "Negative Information: Perspectives and Research Directions." In *Advances in Consumer Research*, vol. 8, edited by Kent B. Monroe, 398–404. Ann Arbor, MI: Association for Consumer Research.

Wolfram S. 1984. "Cellular Automata as Models of Complexity." *Nature* 311 (October): 419–24.

Yale, Laura J., and Mary C. Gilly. 1995. "Dyadic Perceptions in Personal Source Information Search." *Journal of Business Research* 32: 225–37.

CHAPTER SEVEN

The Dual-Market Effect

The Saddle was first reported by Goldenberg, Libai, and Muller (2002b). This chapter is based on that paper and covers other works on the same phenomenon.

In 1985, the prospects of the PC seemed to be in peril. *Business Week* referred to the PC market as "despondent," and sales were reported to be "sluggish" despite an otherwise nonrecessionary US economy. Stephen Wozniak, one of the founders of Apple computers, was quoted as admitting to having "lost his optimism" in the very industry he helped pioneer (Lewis 1985, 142d). This slowdown in sales was all the more unexpected, as it followed two years of rapid growth, each exceeding an average annual growth rate of 50% (Goldenberg, Libai, and Muller 2002b). In hindsight, one may argue that Wozniak's sentiments are symptomatic of a broader phenomenon that affected the entire spectrum of technological goods. In many cases, what was considered a successful introduction of a new product was generally followed by a sharp drop in sales. The temporary drop, or trough, led to subsequent sales that eventually exceeded the initial peak, forming a saddle-like shape in the sales graph. Jacob Goldenberg, Barak Libai, and Eitan Muller (2002b) demonstrated how this phenomenon, which they called the *Saddle effect*, provides important insights about the introduction of new technological products into the market and their adoption patterns. In fact, the Saddle effect is a premier reflection of the main market's resistance to innovations and offers evidence of the uneven distribution of product resistance.

Consider, for example, a small startup firm, which successfully developed a new product. The introduction was successful, and sales begin to show a takeoff pattern. Managers decide to recruit additional personnel and move to a larger and more comfortable facility. Meanwhile, shareholders are interested in their dividends. At the same time, the R&D

department is working on the new product generation, and new investments in recruiting are made. Then, without a warning, sales drop by more than 20%, for at least two years. Such a drop, at this specific timing, could potentially lead to a firm's collapse. As we will see, this is a common scenario, one that firms should be aware of and prepared for.

The Saddle effect discussed in this chapter is a third (following the leapfrogging effect and the diffusion of negative word of mouth) aggregate manifestation of individuals' resistance to innovation (see figure 7.1). We offer several examples, through which we analyze the unique characteristics of a dual market and their implications for product growth. We review several suggestions for mitigating the Saddle effect and its potentially devastating effects for marketers. These recommendations are of particular importance in view of the inevitable nature of resistance and its manifestation, at least as far as technological innovations are concerned.

The Saddle creates many problems for firms when introducing innovations. As we elaborate below, the phenomenon emerges for up to 50% of new technological advances and if not aware of it, firms may face financial problems amounting to financial catastrophes. In fact, in many cases, managers decided to pull the plug on their innovations following initial, deceptive, evidence of product failure. Some of these cases may constitute situations of failing to "cross the chasm" (see Moore 1991), some are potential successes, lacking the foresight that, following the drop in sales, a recovery would have been manifested. The prevalence of this scenario is unclear, but its mere existence suggests that the Saddle is even more prevalent than indicated by our assessments, given that they are made under the assumption of ultimate product success.

The Saddle can be explained by three mechanisms, as discussed in Chandrasekaran and Tellis (2011). The first is related to the structure of a dual market, with a small market that consists of people with low resistance (see chapter 1) and a larger one composed primarily of individuals with moderate-to-high levels of resistance. The Saddle results from the discontinuity in the transition between the two markets.

The second explanation relates to the economic macrostate. A saddle may occur during periods of economic contractions or recessions. There are a few reasons for this. Economic contractions reduce consumers' income and buying power. Consequently, consumers carefully consider their expenses, especially when considering the adoption of innovations. In addition, in unstable economic situations, consumers often lose confidence in their purchasing decisions, making a Saddle even more likely.

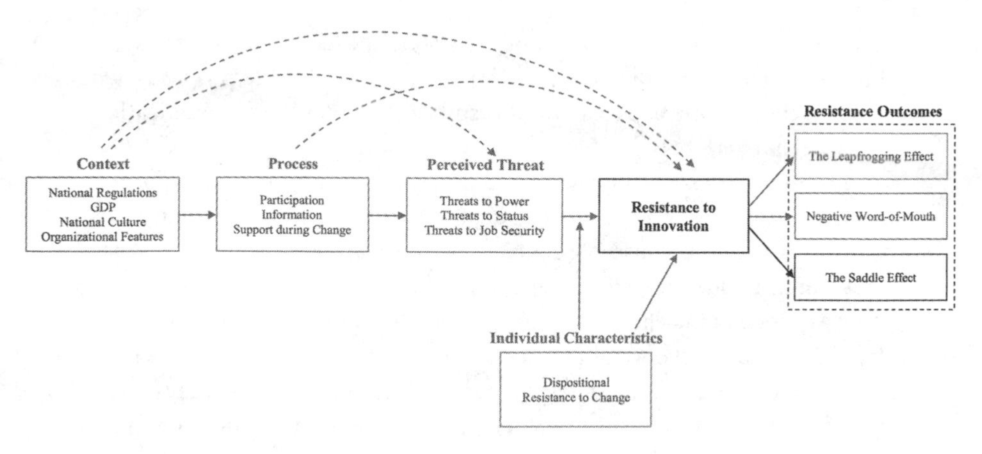

FIGURE 7.1. Resistance and the Saddle effect.

A third explanation for the Saddle becomes relevant in particular during times of important technological advances, in which various product versions compete for market dominance. In these times, many consumers may wait before making significant purchase decisions until the product's standing in the market becomes clear.

Whereas the latter two mechanisms depend on macro-economic changes that may or may not exist, the first mechanism we propose, of a dual market, is relevant for the introduction of any innovation, at any time. It is this mechanism on which we focus in this chapter.

The Saddle as a Dual-Market Phenomenon

One illustration of the significant role of word of mouth in the adoption process is offered in a recent extension of the classic adoption curve—the dual-market phenomenon. It is said to reflect a more accurate representation of new product development by attributing the high failure rate of new products to communications between individual consumers (or the lack thereof). The model, also known as Moore's Chasm Theory (Moore 1991), is supported by a wealth of anecdotal evidence. Ascribing a large number of new product failures to marketers' disregard of an important divergence from classic adoption theory, Geoffrey Moore encourages marketers to recognize that the adoption of a new product critically depends on the market—one of two distinct markets—in which it is received. Rather than a single general market, in which consumers may be differentiated according to their "adopter category," Moore proposes two markets that are formed as a direct result of the inherent communication barrier that exists between early adopters and the main market.

Because of this communication barrier, or chasm, product information from early adopters fails to transfer to main-market consumers, as they form their purchase decisions. This intuitively appealing idea has been confirmed by empirical evidence in Goldenberg, Libai, and Muller (2002a). Beyond the chasm's existence, it is also important to look at sales data to determine the weight of the chasm's impact on product sales relative to many other factors. Indeed, when sales data for innovative products were tracked over time, it became clear that new product adoption takes place in two semiconsecutive loci—two different consumer markets, with distinct consumer attributes. Specifically, using data about a large number of innovative products in the consumer electronics industry, Goldenberg, Libai, and Muller (2002b) found that between 31% to 50% of the cases involved a pattern consisting of an initial peak, giving rise to a trough of sufficient depth and duration to preclude random fluctuations, followed by sales that eventually exceeded the initial peak—a pattern that is termed a "saddle." Golder and Tellis (2004) analyzed new durable product life cycles in thirty categories, and their findings indicate a distinct slowdown of about 15% that occurs on average at approximately 34% penetration. This is a striking finding; it means that the product life-cycle model, for innovative products, consists of two peaks rather than one. Through a hazard model, they showed that the probability of this slowdown is correlated with slower growth of the entire economy, smaller price reductions, and higher penetration rates.

When the inherent differences in the reception of new products by these two markets are sufficiently large, a lag occurs between the adoption patterns of the early market and the main market, creating two distinct sales peaks, rather than the single, classic Bass diffusion pattern. *How does this happen?* At the market level, resistance erodes market growth and sales of innovations by creating two, almost disconnected, markets that challenge conventional thinking of a smooth continuous upward sales curve, generated by the gradual and roughly sequential adoption of an innovation by four adopter groups (we elaborate on these groups below). For the last fifty years, our understanding of innovations and new product adoption has focused on the appeal of such products for potential consumers. As we review below, this approach virtually ignored the role of resistance in determining new product sales.

In the marketing literature, it is common knowledge that not all consumers adopt at the same time because of different levels of *innovativeness*, or openness to innovation. Viewing the market as comprising groups that differ in their level of innovativeness is an approach initially developed by Rogers (1995), who suggested that the adoption pattern of technological goods is the cumulative result of sales among five distinct groups of consumers who adopt at different stages of the product life cycle (Innovators, Early Adopters, Early Majority, Late Majority, and Laggards) and, more importantly, do so for different reasons. Despite the huge popularity that Rogers's classification has gained, the Laggard category, involving the idea that some individuals are inherently reluctant to adopt new products and innovations, has largely been ignored.

In *Crossing the Chasm* (1991), Moore offered a change in the common focus on innovativeness and drew attention to the differences in people's responses to novel products. His approach is grounded in empirical evidence of the Saddle effect, as described above, and specifically, in the discontinuity in sales that creates two distinct peaks in new product sales. In contrast to Rogers, Moore divides the market into two groups—visionaries (the minority) and pragmatists (the majority). Visionaries (or Innovators) are the early market, those adopters who eagerly seek innovations and are attracted by the technological advancement itself, almost regardless of its functionality. Main-market pragmatists, however, are cautious and attribute greater importance to functionality. They need more time to make an adoption decision, which gives marketers time to introduce improvements and adaptations to the original version of their product. According to Moore, for a firm to make the transition from initial, yet limited, market success, to mass market success, it must "cross the chasm" between these distinct groups. To succeed, firms must be attentive to and address the distinct needs of both types of adopters rather than assume that the same product will meet both groups' needs and desires. The concerns of main-market consumers require firms to pay more attention to product costs, compatibility, simplicity, and reliability. Note that Moore's main assumption about the reasons for resistance is that something is wrong, or at least not entirely right, with the innovation. He assumes that people resist innovations for pragmatic reasons (see chapter 2).

Contrarily, Innovators, according to Moore, are distinct in their nature from the more resistant main market. They are like David, whom we described in the introduction to chapter 5; they love innovations, are attracted to breakthrough ideas, and are willing to spend time and money to explore them. One could say that they show no resistance to innovations.

The following story illustrates a fundamental difference between early- and main-market adopters. Several years ago, a colleague of ours (Barak Libai from Israel's Inter-Disciplinary Center in Herzliya) possessed one of the first webcams available on the market. It was a cumbersome device. Internet connection speeds were much slower at the time, making the transmission of media over the web a weary process. The webcam nevertheless captivated Barak, who was not overly concerned about its limited functionality. When he showed Jacob his webcam, Jacob immediately thought about how much he could gain if he could share his sketches and formulas with his colleagues through a webcam without having to send them over the web—allowing him to save time and effort. He purchased one right away. Not long afterward, Barak's webcam broke. He did not hurry to have it repaired or replaced. His fascination with the webcam had by then waned, and he was already on the lookout for his next gadget. Barak was enchanted with the technology and its innovative nature and was much less interested in the pragmatic opportunities it offered. Our distinct approaches represent the difference between visionaries and pragmatists. While the innovator is fascinated by the innovation per se, the main-market pragmatist views it from an instrumental perspective. The innovator purchases to acquire the innovation, whereas the main-market adopter is interested in its functionality. As we discussed in chapter 1, marked dispositional differences likely exist between these two groups of individuals.

Moore's visionary category includes both Rogers's Innovator and early adopter categories, which together are estimated to comprise between 13% and 19% of the market. Early adopters are often quite distinct from Innovators, such as Barak in that most of them are pragmatists, much like the main market, and are interested in the innovation for its advantages rather than for its mere newness. Unlike the main-market consumer, however, its newness does not intimidate the early adopter, who exhibits little, if any, resistance to innovations. Contrarily, most consumers belong to the *main market* and exhibit some degree of resistance to the innovation (people like Aaron, whom we described earlier, in chapter 5).

This is a coarse-grain view of the market segments, also called a dual-market approach—a small-size market that is fast to respond to innovations, along with a large market that is slow in its response, reluctant to make changes and showing resistance to innovations. In part I of the book, we explained the reasons consumers in this latter market have for their resistance: because of their personality (chapter 1), the context or timing

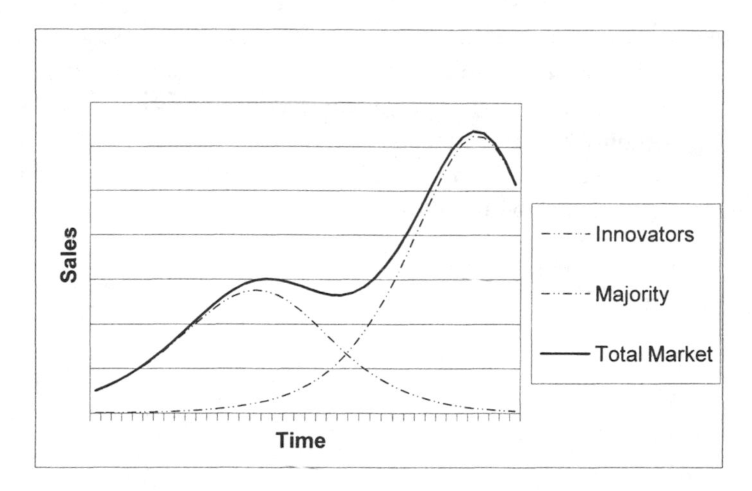

FIGURE 7.2. Dual market: the Saddle mechanism. Adapted from figure 3a in Goldenberg, Libai, and Muller (2002b).

of the innovation's introduction (chapter 4), the manner in which it was introduced (chapter 3), or, ultimately, because it is perceived as a threat (chapter 2), not justified by its added value.

Figure 7.2 is a graphic illustration of how the dual market relates to the Saddle phenomenon (more precise parameters of the relationship between the two markets are specified below). The two markets begin at the same point in time and, although not isolated from each other, each has its own market attributes and potential. The diffusion and growth of the product are plotted and calculated separately for each market.

The prevalence of the Saddle effect is surprisingly high. Even the sales curves of successful new products show a Saddle. Examining the adoption rates of three consumer electronics products (as an illustration, see figure 7.3a–c for the cases of personal computers, VCRs, and cordless telephones, respectively) reveals a perplexing phenomenon. In all three cases, products reach an initial peak in sales shortly after penetrating the market, then experiencing an abrupt but significant decline in sales, only to be followed by a second peak that significantly exceeds the initial peak (figure 7.3a–c; the dependent variable is the number of unit sales of these products in the US). This adoption pattern is the Saddle effect. As we explained above, it is not an anomaly. It occurs regularly and its existence is supported by a large corpus of data of thirty-two innovations compiled by the Electronic Industries Alliance (presented briefly below). In approximately one-third of these cases, new innovations lose significant ground before regaining a strong sales profile. Note that this temporary postlaunch slump in sales cannot be attributed to a recessionary economy (the US economy was by no means recessionary throughout the 1980s) (Goldenberg et al. 2006). Rather, it reflects the resistance that at least part of the market exhibits toward innovations.

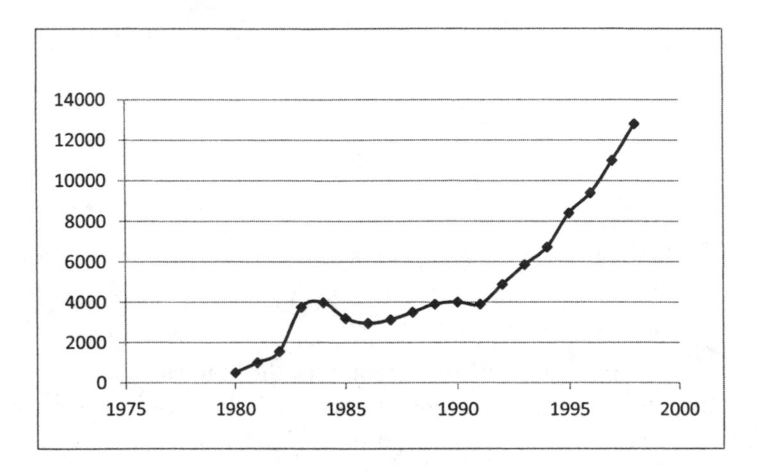

FIGURE 7.3A. Saddle in personal computers. Adapted from figure 1a in Goldenberg, Libai, and Muller (2002b).

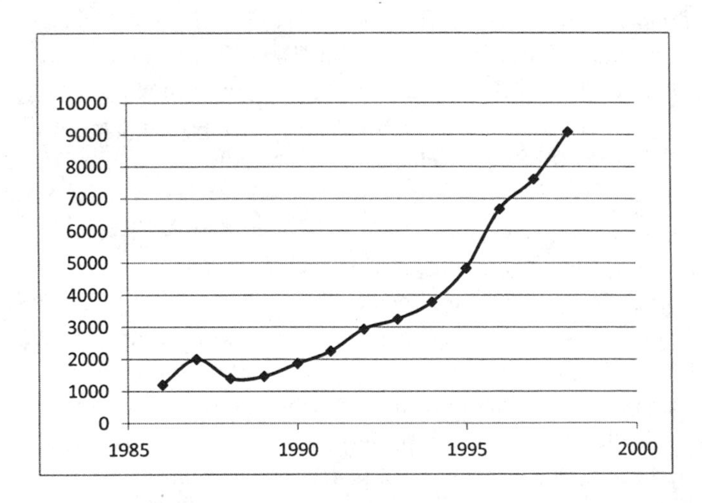

FIGURE 7.3B. Saddle in VCR decks with stereo. Adapted from figure 1b in Goldenberg, Libai, and Muller (2002b).

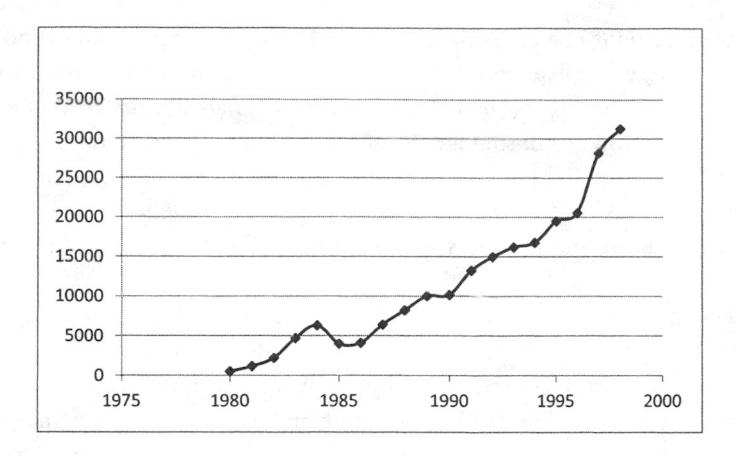

FIGURE 7.3C. Saddle in cordless phones. Adapted from figure 1c in Goldenberg, Libai, and Muller (2002b).

As can be seen in these figures, in each case, a local maximum in sales occurs shortly after product launch and takeoff (in 1984 for PCs, 1987 for stereo VCRs, and 1984 for cordless phones). Subsequent to this local maximum, a considerable drop in sales occurs over a period of a few years (a drop of 30% for PCs over a period of seven years, 30% for VCRs over a period of three years, and 35.5% for cordless phones over a period of three years).

Another interesting case is that of the citizens band (CB) radio market. A CB radio is a two-way communications radio that any civilian (as distinguished from members of the police force) can use to communicate with other CB radio operators. The beginning of the CB radio industry is traced to 1958, when the FCC formed the basis for the CB as it is known today. The early diffusion of CBs peaked at around 1963, and sales generally declined thereafter. Growth resumed in the early 1970s and increased rapidly in the mid-1970s, creating what was described in the popular press as a "market explosion." By 1976, the CB radio was considered the "consumer electronics star performance of the year" (Electronic Industries Association 1976), and over five million CB license applications were filed in 1977 alone. After 1977, demand declined, yet remained considerably higher than in the early years.

According to industry literature, there was a noticeable difference between the average CB user in the 1960s and '70s. Users in the '60s were considered more serious, who either needed the CB for work (such as farmers or boat owners who needed to keep in touch with their home base) or were deeply involved hobbyists, interested in radio transmission. Contrarily, in the 1970s, the CB radio became "mainstream," in a market composed of many consumers with no record of, or high need for, radio communication. The product created a widespread hobby, as equipment became available through mainstream consumer outlets. By the end of the 1970s, however, many CB enthusiasts had left the market, turning to other hobbies.

The increase in demand for CB radios in the early 1970s (that appeared only after a decrease in the late 1960s) is intriguing, as no major change in the product itself or its price can account for it. The Electronic Industries Association (EIA), the main industry data source that followed the CB radio's diffusion, attributed the change to communication factors, suggesting that the change resulted from "increased awareness of the availability and utility of citizens radio," which occurred around 1972 (Electronic Industries Association 1976). Although it is unclear whether this increased awareness should be traced to the word-of-mouth communication of the 1960s market, or to product advertising in the mass media, it is clear that a large mainstream market emerged in the early 1970s to create a clear and strong Saddle effect.

That sales development over time is marked by fluctuations is not a new insight. As mentioned above, Peter Golder and Gerard Tellis (2004) convincingly demonstrated that the life cycle of a product is not marked by monotonic and progressive growth, as is usually taught in marketing management courses. Analyzing a varied set of products over a period of almost a century, they show that product sales development over time is marked by intervals of rapid growth (45% over the first eight years from product launch) and manifest slowdowns. The Saddle effect is a special case of such fluctuations, based on the nonuniform distribution of resistance to innovations in the market.

The Saddle provides yet another piece of evidence for the existence of resistance to innovation. It clearly demonstrates that most people are not as captivated with innovations as are the early adopters. For the main market, resistance to change is a natural part of the normal frame of mind. Main-market consumers are more likely to resist innovations until it is expedient or until they are forced to adopt (for example, when changes in the market make extant options obsolete). Although in reality, early adopters and main-market consumers are both interested in the novelty and functionality of innovations, they differ markedly in the weights they assign to each. As noted above, early adopters are attracted to innovation per se, and they heavily discount functionality. Main-market adopters, contrarily, hold their potential new products to a higher standard of functionality. In addition, the greater the adjustments they are required to make to accommodate the innovation (see chapter 2), the greater their resistance to the innovation.

Main-market consumers recognize that adopting a technological innovation entails not only the need to learn new habits. Adoption entails additional costs and risks as well, as a necessary step for progress to occur. In his March 16, 2010, column in Slate, Farhad Manjoo describes the dilemma of the resistant main marketer:

> There's certainly ample historical evidence that buying tech early isn't smart. The first generations of new devices are usually more expensive, more buggy, and offer fewer features than later generations. . . . If a better, cheaper version is always around the corner, why should you ever hurry? The problem with such a philosophy is that, at its extreme, it results in paralysis; if you're always waiting for next year's device, you'll never buy anything. At some point you've got to accept that tomorrow's cheaper, more advanced versions aren't around today—and then go to the store and see what's available in the present.

Some main-market consumers resist adoption until obsolescence has eliminated all alternatives (see chapter 5). Manufacturers are responding to such concerns and are reducing the risks associated with obsolescence by offering updates and functionality to older products, which in many cases can be downloaded from the Web.

Main-market consumers' intuitive preference to hold off on adoption, at least with respect to technological innovations, is also justified by the fact that the benefits of many technological innovations are based on their presence within a network (termed *network externality*). In such cases, as the perceived cost of adoption declines, product adoption rates increase. Innovations such as Blu-ray players and e-book readers are examples of innovations that become more useful in direct proportion to the number of existing users, given that as adoption rates increase, so will the availability of relevant products (such as Blu-ray disks and e-books) that could be used with the innovation. For main marketers who put a high premium on functionality, this is yet one more reason to accede to their natural resistance and wait for a later model.

Measuring the Saddle Effect

Our next aim is to assess how common the Saddle effect really is. To do so, we begin with a clear definition of a Saddle that allows us to disregard local fluctuations in new product sales and noise in sales data. We can then isolate cases of product sales that show a Saddle effect.

Following Goldenberg, Libai, and Muller (2002b), let d be the depth of the Saddle, measured as the sales difference from the initial peak to the lowest point of the Saddle. Let w be the duration of the Saddle, measured as the time elapsing from the initial peak (T_s) to the recovery of sales levels. Clearly, to distinguish between a Saddle and random perturbations in the market growth pattern, some threshold for w and d should be determined. Admittedly, the specification of these minimal sizes is arbitrary, yet Goldenberg, Libai, and Muller (2002b) selected the following conservative measures for defining a Saddle as a trough following an initial peak in sales, reaching a depth of at least 20% of the peak (10%, if we use a more relaxed definition of a Saddle, see below), lasting at least two years, followed by sales that ultimately exceed the initial peak.

Denoting h as the initial peak sales level, and d^* as the relative depth, that is, d / h, Goldenberg, Libai, and Muller (2002b) then define the following conditions for the occurrence of a Saddle (see figure 7.4).

$$w \geq 2 \; and \; d^* = d / h \geq \begin{cases} 20\% & strict\ definition \\ 10\% & relaxed\ definition \end{cases}$$

Chandrasekaran and Tellis (2011) report that in their sample the Saddle takes place, on average, nine years post-takeoff, at a mean penetration of 30%, lasting eight years with a 29% drop in sales at its depth. All three sales patterns in figure 7.3 satisfy this strict definition. As we elaborate below, using the Electronic Industries Alliance data set mentioned above, Goldenberg, Libai, and Muller (2002b) showed that the occurrence of a Saddle in new product growth may occur in one of every two new products, and is particularly conspicuous in technology-oriented markets, where it is widely recognized as a prevalent sales development feature.

Table 7.1 lists descriptive data of the sample investigated. The original data set includes sixty-two innovations, primarily in the consumer electronics industry. After eliminating all cases that contained few points of data or that did not contain data on unit sales, we remained with a sample

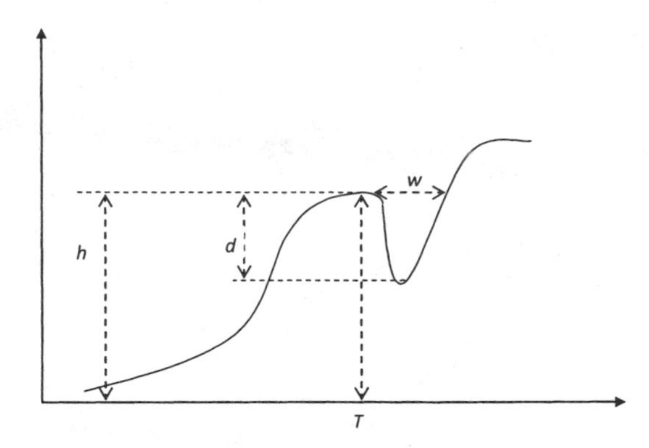

FIGURE 7.4. Structure of a Saddle. Adapted from figure 2 in Goldenberg, Libai, and Muller (2002b). $d^* = d / h$ = Saddle relative depth; w = Saddle duration; T_s = Time of beginning of saddle.

of thirty-two valid cases. Of the thirty-two cases, ten were found to have a duration and relative depth that satisfied our strict definition (a relative depth of at least 20% and a minimum duration of two years), and one case satisfies our relaxed definition. However, if the constraints are further relaxed to include Saddles with a minimum duration of one year *or* a minimum relative depth of 10%, the percentage of cases with a Saddle increases to 50% (these cases are marked with an asterisk [*] in the following table).

The prevalence of the Saddle has also been estimated in other research. A number of recent articles empirically document the phenomenon of a trough in sales. For example, Golder and Tellis (2004) define a "slowdown" (or start of the Saddle) as the end of the growth phase in a product's life cycle and find that the slowdown occurs in twenty-two of the cases in a sample of twenty-three products within the United States. Chandrasekaran and Tellis (2011) empirically analyzed historical sales data from ten products across nineteen countries and found that the Saddle exists in 148 (78%) of the cases. In these data as well, the Saddle indeed seems to be widely prevalent.

Yet, at least some of these assessments of the Saddle's prevalence may be underestimates. As noted above, Goldenberg, Libai, and Muller (2002b) assessed that the Saddle exists in up to 50% of all *successful* new product introductions. In some cases, however, the Saddle may be so deep,

TABLE 7.1 **Distribution of Saddles in sample data**

Product description	Saddle	Relative depth	Duration (in years)
Answering machines			
Blank audio cassettes	yes*	6.6	2
Blank floppy discs			
Camcorders	yes*	5	2
CD players			
Cellular phones			
Color TV	yes	35.6	4
Color TV with stereo			
Compact audio systems	yes	53.5	11
Cordless phones	yes	36.5	2
DBS satellite	yes*	21.4	1
Digital corded phones	yes	36.7	7
Fax machines			
Fax modems			
Home radios	yes	33.6	6
Laser disc players	yes*	11.8	1
LCD monochrome TV			
LCD color TV			
Monochrome TV	yes	27.8	3
Personal computers	yes	25.8	5
PC monitors	yes	25.8	5
PC printers	yes	25.8	5
Portable CD equipment			
Portable tape and radio			
Projection TV	yes*	12.3	1
Rack audio systems			
TV / VCR combination			
VCR decks	yes*	18.7	5
VCR decks with stereo	yes	30	3
Video cassettes			
Video cassette players			
Word processors			
Percent saddle (strict definition)	34.4%		
Percent saddle (relaxed definition)	50%		
Average (strict definition)		31.8%	5.1 years
Average (relaxed definition)		25.4%	3.9 years

and occur so soon after product launch, that the new product is withdrawn from the market altogether (and thus never becomes a success). These cases are obviously absent from Goldenberg and colleague's calculations, and so the 50% estimation may in fact be a conservative estimate. Such a high prevalence may require an entirely different view of the Saddle effect. The phenomenon is not external to the product, but part of the product's natural life cycle.

Several researchers have explored how the Saddle develops and how product sales transition from the early market to the main market. In line with Moore's (1991) distinction between visionaries and pragmatists, these researchers show that the chasm—the duration of the drop in product sales after an initial peak—is caused by differences between the two distinct adoption patterns attributed to the early and main markets.

For example, in their 2006 paper, "When Does the Majority Become a Majority? Empirical Analysis of the Time at Which Main Market Adopters Purchases the Bulk of Our Sales," Eitan Muller and Guy Yogev analyze the chasm by asking how much time elapses until main market adopters outnumber early market adopters. Their study focuses on the transition period, when early adopter dominance of sales is transformed into main-market dominance. They use sales data of twenty-six products, 75% of which showed two distinct sales trajectories conforming to the dual-market structure. Defining CD-time (Change of Dominance Time) as the number of years from product launch until the number of main-market adopters exceeds that of early-market adopters, they found that the median CD-time was 7.5 years. Over 75% of the products they studied exhibited a shift in dominance within five to ten years of product launch, and found that the average adoption rate at CD-time is 16%. That is, after 16% of the consumers have adopted the product, sales became dominated by main-market consumers. Sixteen percent is a familiar number; as it turns out, Rogers, who used standard deviations to divide the population into the five consumer categories, came up with the same value (16%) to represent the size of the early market.

In their paper, "When Giving Some Away Makes Sense to Jump-Start the Diffusion Process," Donald Lehmann and Mercedes Esteban-Bravo (2006) propose subsidizing (or "seeding") a small number of influential early adopters to accelerate diffusion and thus the product's overall adoption rate. Lehmann and Esteban-Bravo demonstrate the expediency of granting incentives to a specific group of influential consumers (also known as "influentials") whose purchase decisions are believed to trigger a strong chain effect, influencing many others to adopt. The authors suggest that the product adoption rate is most likely to grow when the market is "seeded" soon after product launch, before the drop in sales develops.

Golder and Tellis (2004) as well as Chandrasekaran and Tellis (2006) also showed that the Saddle phenomenon could be explained using the informational cascade theory, where *informational cascade* is defined as individuals' tendency to adopt a behavior based on the value of the signal they derive from previous adopters' behavior (Golder and Tellis 2004). Correspondingly, Christophe Van de Bulte and Yogesh Joshi (2007) provide an overview of the dual-market phenomenon and its relation to the diffusion

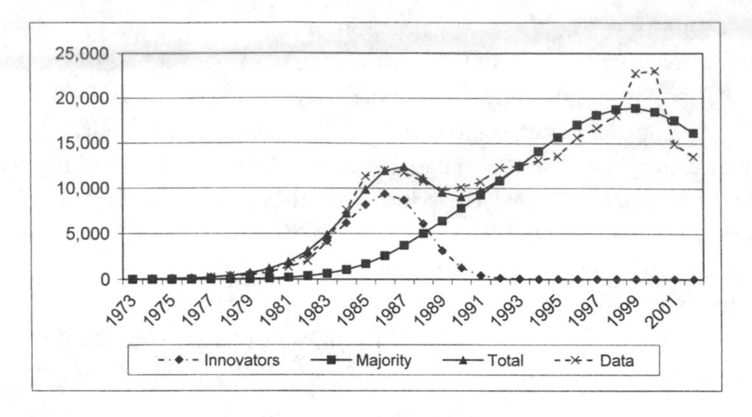

FIGURE 7.5. Sales of VCRs in the United States (in thousands of units).

of innovation. They show that a drop in diffusion is one of several types of diffusion curves that may appear in a dual-market context, and that the decline in imitators' adoption need not be monotonous. In contrast to Moore, Van de Bulte and Joshi's analysis demonstrates that product modifications may not be necessary for gaining traction among main-market adopters.

Finally, as noted above, Chandrasekaran and Tellis (2011) found evidence for a Saddle in 148 of 190 cases (ten products in each of nineteen countries). They indicate that the dual market and technological cycle mechanisms are responsible for the Saddle in information/entertainment products, but found no evidence for the dual-market mechanism in kitchen/laundry products. It is plausible that in this latter product category, which is less visible and socially significant than information/entertainment products, the word of mouth of the two segments (early adopters and majority) are less discontinuous, which is why this mechanism does not account for the Saddle.

Timing has a critical effect on the development of the two markets and the emergence of a Saddle. In figure 7.5, the penetration of the VCR is presented along with the two-market subpenetration. The two-market penetration curve was derived using the Goldenberg, Libai, and Muller (2002b) model. A Saddle phenomenon is observed if the growth in main-market sales begins late. As can be seen in figure 7.5, however, the main market takes off shortly after sales in the early-market peak. If the two markets either take off simultaneously, or, as is the case for VCRs, where the main-market takeoff begins slightly after early-market sales peak, then a Saddle does not emerge. In the VCR case, we see a small "deformation" in the sales curve, hinting at a potential Saddle that might have emerged, had the gap between the two markets been larger.

Debunking the "First Mover Advantage" Myth

Main-market resistance is also related to an interesting observation that the "first mover advantage" (also known as pioneer advantage) may, in fact, be a myth. For a long time, a simple and conclusive relationship was assumed between pioneering (being the first to commercialize a product in the market) and market leadership (the ability to persistently dominate the specific market category). With this reasoning, marketing could be seemingly reduced to a single basic maxim: being first equals success, with all other considerations being of marginal importance.

Closer scientific scrutiny, however, revealed that the assumption underlying this view was misguided. In a series of essays and, more pronouncedly, in a book titled *Will and Vision*, Tellis and Golder (2001) discredited the belief that pioneering ensures market leadership and established the role of several other factors that account for market leadership, such as vision and will. They demonstrate that the pioneer-leader equation has become so ingrained among marketing scholars and practitioners that it persists, even when evidence suggests otherwise. Apple, for instance, is often considered the founder of the personal computer, whereas Micro Instrumen tation and Telemetry Systems (MITS) sold the Altair in 1975, preceding Apple by a year (Tellis and Golder 2001). It seems that the real pioneers may fade from public memory as soon as they lose control of the market, whereas market leaders, who may have been only second or even third to enter the market, become firmly fixed in the people's minds as the pioneers. The same notion is illustrated in the McDonald's case. The first McDonald's branch was opened in California and managed by the founding brothers who showed little, if any, interest in commercial expansion. It was Ray Kroc, a milkshake machine franchiser, who saw the potential in expanding the brothers' concept and who is mistakenly considered the founder of McDonald's (Tellis and Golder 2001).

As it turns out, market pioneers rarely endure as market leaders. Analyzing data compiled over ten years, Golder and Tellis (1993) and Tellis and Golder (2001) conclude that pioneering is neither a necessary nor sufficient condition for enduring success. More important are traits such as vision and will. In the laser printer category, for example, IBM and Xerox were first in the market, now dominated by HP. In the Internet browser category, Microsoft's Internet Explorer is usually perceived as one of the pioneers, even though it was introduced only after several other browsers, one of which was NCSA Mosaic, although it too was not the very first browser. We provide here a partial chronological listing of some of the influential early web browsers that advanced the state of the art:

World Wide Web (1991, the first web browser, invented by Tim Berners-Lee)
Erwise (1992)
ViolaWWW (1992)
Midas (1992)
Samba (1992)
Lynx (1992)
Mosaic (1993)
Arena (1993)
Cello (1993)
Internet in a box (1994)
Navipress (1994)
Netscape (1994)
Mozilla (1994)
Internet Explorer (1995)

Despite this historical fact, Microsoft's Internet Explorer enjoys almost uncontested market hegemony. In fact, according to the statistics that Tellis and Golder present, pioneers are actually more likely to *not* dominate the market (market pioneers are leaders in only a small percentage of the categories surveyed in their study). Another interesting example is Google, which, although a dominant player (and firm), is an amazingly new search engine. Here is a partial chronological list of the most dominant search engines until the rise of Google (taken from http://searchenginewatch.com/2175241):

Lycos (1994; reborn 1999)
WebCrawler (1994; reborn 2001)
Yahoo (1994; reborn 2002)
Open Text (1995–97)
Magellan (1995–2001)
Infoseek (1995–2001)

Excite (1995; reborn 2001)
AltaVista (1995–)
HotBot (1996; reborn 2002)
LookSmart (1996–)
Snap (1997–2001)
AOL Search (1997–)
Direct Hit (1998–2002)
Ask Jeeves (1998; reborn 2002)
MSN Search (1998–)
Overture (1998–)
Google (1998–)
AllTheWeb (1999–)
Teoma (2000–)
WiseNut (2001–)

Although Tellis and Golder dispelled beliefs about the alleged pioneer-leader link, they did not explain why this compelling intuition is so wrong, and why marketing pioneers lose their edge despite having the early start. Individuals' resistance to innovations may be the explanation to these issues as well. Recall that the nonuniform distribution of resistance to innovation creates different groups of consumers, each with its distinct adoption pattern. At an aggregate level, these differences in adoption patterns appear as the Saddle. Because early adopters are more interested in the novelty of the product than its functionality, pioneers can afford to launch novel products that are not fully developed, but whose features are geared toward those seeking innovation for innovation's sake. Firms that are second or third to launch their products typically have a more mature product with better functionality and fewer problems. Accordingly, these products are more suitable for main-market adopters.

Seen this way, a pioneer's loss of market leadership is not so much a failure to dominate the market as it is a success in controlling a specific market segment. The problem is that it is not this segment that matters most. Pioneers who attract the early market but fail to make the necessary modifications to appeal to the main market will not be able to translate their initial success to main-market domination. And a product can't succeed if it fails to penetrate the main market. Following Tellis and Golder, we argue that the pioneer-leader equation depends on the firm's ability to perfect its product and cross the chasm—by offering a product that is both novel and functional.

Determinants of Saddles' Depth and Duration

When should we expect a Saddle, and how deep will it be? Some of the properties of the dual market provide hints for addressing this question.

The following conclusions of the main Saddle determinants are taken from Goldenberg, Libai, and Muller (2002b).

THE EFFECT OF EARLY MARKET SIZE. Given the existence of a Saddle, the larger the early market, ceteris paribus, the larger the Saddle's relative depth and duration. When the main market is slow enough to adopt so as to create a Saddle, an increase in the initial peak leads to a corresponding increase in Saddle parameters. The main effect is indicated in the Saddle's depth rather than its duration.

THE MAIN MARKET EFFECT. Both marketing efforts and within-market communications parameters (communication within each of the two submarkets) have a strong negative impact on the relative depth and duration of the Saddle. The rationale for this lies in the fact that, to the extent that these two increase, main-market adopters have a greater impact early in the product adoption process. Accelerated main-market growth decreases Saddle parameters, in terms of both relative depth and duration.

THE EFFECT OF THE CROSS-MARKET COMMUNICATIONS PARAMETER. Cross-market communication increases Saddle prevalence, as long as it is weaker than within-market communication.

THE EFFECT OF WITHIN-EARLY-MARKET COMMUNICATIONS ON SADDLE DEPTH. Of all parameters, early market word of mouth has the largest effect on relative Saddle depth.

THE EFFECT OF WITHIN-EARLY-MARKET COMMUNICATIONS ON SADDLE DURATION. Counterintuitively, high values of *early market communication* significantly reduce Saddle duration. When this communication is substantial, one could expect the early market to adopt at a faster rate, creating a larger window for the more reluctant main market consumers to adopt. Accordingly, Saddle duration would be expected to increase when within-early-market communications accelerate early-market growth (especially in the case of a reluctant main market). Findings indicate, however, that in such a situation, Saddle duration actually *decreases*. This is because the effect of communication acceleration is not linear. Recall that with small values of early-market communication, Saddle formation is possible only when the value of main-market communication is also small. In this subrange of parameters, a small increase in early-market communications increases the pool of early market buyers, allowing cross-market communications to activate the main market at an earlier point in time. This indirect effect induces a reduction in the Saddle's duration.

THE EFFECT OF EARLY-MARKET MARKETING EFFORTS. Another seemingly counterintuitive result is the negative impact of marketing efforts directed at the early market on the relative depth and duration of the Saddle. To understand this result, note that such an impact is contingent on there being a Saddle in the first place. Given the existence of a Saddle, weak marketing efforts can instigate substantial internal early-market communications. This, however, also implies that early-market growth is steep (Rogers 1995; Mahajan, Muller, and Srivastava 1990), leading, in turn, to a high initial peak. As marketing efforts increase, the range of the early-market communication expands to include lower values, resulting in a slower rise to a lower initial peak and a shallower Saddle. This will not occur when marketing efforts are substantial to begin with. This also explains the effect on Saddle duration; to the extent that the initial peak is higher, a longer duration is required for sales to regain the initial peak level.

THE INTERPLAY BETWEEN EARLY-MARKET SIZE AND THE DELAY IN MAIN-MARKET ADOPTION. Of all main and significant effects, early-market size has the least impact on Saddle duration. The reason is that, ceteris paribus, a large early market delays the initial peak, requiring more time for most early-market members to adopt. This delay allows the main market to enter the adoption process before the early market completes its adoption process.

Avoiding the Saddle Hazard

From a managerial perspective, the Saddle phenomenon warrants attention because a significant and unexpected decline in sales, at the relatively early stages of a product's life cycle, may erroneously cast doubt on product viability. Identifying the conditions required for the formation of a Saddle may therefore prevent the premature withdrawal of new products. This is especially true for high-tech and similar innovative products, because firms typically continue R&D and product improvements after market launch, increasing their vulnerability to early sales fluctuations.

The Saddle effect is dangerous because it occurs at a critical stage, usually when penetration reaches 16%. This point in the product life cycle, known as takeoff, is when most managers expect to finally see results that will allow them to forecast profits. This is also the stage in which managers invest additional resources in rising products, planning the next product generation. As we demonstrated, however, this is also when the Saddle usually begins to form. When it does, reserved optimism fades as growth is replaced by a slowdown. Reduced cash-flow generation combined with a steady decline in sales poses a genuine threat to the future of many products. As a result, new products may be withdrawn from the market. If, however, the drop in sales is recognized as the beginning of the Saddle rather than the product's ultimate descent, managers and marketers may refrain from terminating products, offering them a chance to regain popularity.

The more we understand the Saddle phenomenon, the more we are able to adapt to it, for example, by releasing cautious profit forecasts, preparing for future cash-flow shortages, or designing future investments. In fact, in some cases, it is not difficult to estimate the size of the saddle even as early as at the product's takeoff. The more pronounced the difference between early- and main-market resistance levels (information on which can be obtained through market research), the easier prediction becomes.

Adapting to the Saddle can take on different forms. Two specific methods are intuitively appealing but are more theoretical than practical. One option is to disregard the early market altogether when planning the product's marketing strategy. The early market is sure to purchase the product regardless of marketers' strategy, and so marketers can devote their attention to the main market, which matters most. Yet attention to the early market is required given that it is a source of pertinent feedback that marketers use to fine-tune products' features, maximize their compatibility, and thus increase their appeal to the main market. Early market adopters act as firms' beta testers. Not only are these consumers eager to adopt the innovation despite its flaws; they also help identify failures and enjoy proposing potential improvements to product developers. Given main-market consumers' low tolerance for problems with their products, this early-market testing ground substantially improves the chances of overcoming main-market resistance.

A second option is to apply a radical marketing expense allocation model that dramatically reduces the costs of marketing at the initial introduction stage and focuses on later stages, once the product is ready for the main market. This view stands in stark opposition to common advertising models that call for an initial marketing blitzkrieg followed by steady maintenance. This option, however, is also easier said than done, especially in view of inter- and extraorganizational factors, such as the impetus to spend rather than hoard allocated funds, and current relationship models with advertising agencies, which give advertising agencies little incentive to support a deferred spending schedule.

Another difficulty that may emerge in adapting to the saddle is that of transferring experience from one innovation to another. We have stated earlier that resistance is not distributed uniformly across consumers. Furthermore, even the same consumer will respond differently to different products. Some product categories are more susceptible to a chasm than others. Radical innovations—products that are particularly innovative, requiring consumers to alter basic behaviors and tastes—are more likely to generate a significant Saddle effect than less radical innovations.

Saddle Characteristics Depend Not Only on the Product but Also on the Firm

The degree of disconnection between the early and main markets is strongly influenced by firms' specific product and market strategies. As a result, not all firms in the same market experience the Saddle to the same degree. The second author (Jacob) recalls a lunch meeting with Henk Speijer, a devoted Philips manager and, at the time, head of the company's development department. In a private conversation, I (Jacob) told him, based on data I had, that the DVD, one of the company's most cherished products, was experiencing a Saddle. Henk responded with a smile and was quick to cut me off. He said that I was an academic, and that I clearly lack the insight of an insider, a genuine "industry expert." "Trust me," he said, "a Saddle is an interesting phenomenon, but it doesn't exist in DVDs." Paradoxically, we were both right. I was right in claiming that certain brands (and certain models) of the DVD were experiencing the beginning of a Saddle, and Henk was right in arguing that they, at Philips, weren't. The thing is, we were each looking at different data: I was basing my assessment on the US sales of a subset of brands, whereas Henk was referring only to sales of the Philips DVD player. Interestingly, while total DVD sales were experiencing an overall decline, Philips's DVD sales remained stable, which shows that the emergence of a Saddle, and Saddle characteristics, depend not only on the product, but also on the specific firm. Some firms may leap over the trough altogether, whereas others may experience such a deep trough that they find too difficult to bound and ultimately go out of business.

What we have yet to reveal is that Philips's success in avoiding the global Saddle was not a coincidence, but rather a result of highly professional work. When DVDs emerged in 1997, VCRs had enjoyed popular mass-market consumer success for over twenty years. VCRs looked like typical stereo components—rectangular black (or silver) boxes with lots of buttons on the front panel and the typical display that presented the time and current device mode. At some point, VCRs had so many features that consumers had a hard time using them. Even setting the time became complicated.

Relative to the VCR cassette, however, the DVD had several substantial advantages. For starters, DVDs were much thinner, a mere 3/64 of an inch thick, compared to the bulky magnetic 1-inch cassettes. DVDs loaded faster and were easier to operate. With a DVD player, one could skip to different "chapters" of a movie without having to run through the entire tape. DVDs were also easier to store, easier to play, easier to manufacture, and easier to sell. With the introduction of the DVD player, manufactures had a golden opportunity to capitalize on the disk's features and introduce a radically different player that would overcome many of the mounting disadvantages of VCRs.

Surprisingly, they did not. Consider the beginning of the DVD craze, around 1998. Despite the tremendous advantages DVDs had over VCRs, companies were not differentiating DVD players' appearance from that of VCRs, with a brand's DVD player looking almost the same as its VCRs in size, shape, look, and feel. Whereas people were accustomed to the VCR and knew how to use its basic functions, the DVD was new, and the multiple functions and buttons had become overly technical, ultimately yielding resistance among the market majority.

The team at Philips was looking for a way to differentiate their players from those of their competitors. Although they probably weren't aware of the Saddle phenomenon, they did notice the oddity involved in producing

DVD players that replicated the appearance of VCRs, despite the vast differences between the DVD disk and the VCR cassette. Together with an innovation-support firm, called SIT (Systematic Inventive Thinking), Philips systematically designed a DVD player that would elicit less resistance. One by one, Philips removed components and functions while maintaining the player's basic features intact. It started by removing the front panel buttons, at which point consumers had a hard time recognizing the device as a DVD player, given that, at that time, no one had seen a DVD player without buttons.

The turning point was when one of the designers on the Philips team suggested making the device substantially slimmer than the typical VCR. After all, a thick device, with ample space for placing buttons, was no longer necessary given that most of the player's functionality could be controlled with the remote control. Buttons on the device were moved to the side and back of the player, just in case owners lost the remote. Aesthetically, this gave players a much sleeker appearance, with the added benefit that the device would fit compartments that were much narrower. But the real advantage was that the player was much less intimidating than were extant players. Other players were resisted because of the perceived threats they presented (see chapter 2), and the new player seemed much simpler and easier to use. Following this approach, Philips engineers removed the large LCD screen from the front panel, which allowed them to make the unit even thinner, resulting in Philips engineering the thinnest DVD player in the industry. It was dubbed the Slimline, and it wasn't long before the entire DVD industry adopted the Slimline as the dominant design. Although not with deliberate awareness of its consequences, it was this approach by Philips, with a particular focus on the resistance that extant designs elicited, that prevented an industry-level Saddle.

Conclusion: The Case of Apple

As noted above, alongside the dangers generated by a misleading Saddle, there are several advantages to the dual-market phenomenon. The Saddle is often a necessary stage for the successful introduction of an innovative product, allowing for simultaneous introduction and improvement. As we argue above, transferring responsibility for product adoption and take-off to the main market is an expensive decision, as is luring the hesitant main market, which would be far more costly than investing in the early market. In addition, by "listening to their voice," firms solicit feedback from the early market as important input for further R&D and product improvements. Firms may therefore prefer to optimize rather than eliminate the Saddle. They could manipulate Saddle size and timing by allocating resources and deciding on appropriate marketing strategies based on the Saddle's predicted appearance. The Saddle is part of the innovation life cycle, and when its existence is acknowledged while plans are made and resources are managed, it does not necessarily present a problem. Nevertheless, some firms have found a way to avoid Saddles altogether. One of them is Apple, in the second Jobs era.

Gordon Goble, on the Digital Trends website, writes, "Apple . . . has concocted and released some of the most imaginative, groundbreaking, and iconic products of the digital age. It has continually set the tone for style and ease of use, and has harmonized the relationship between man and computer more efficiently than any other company on the planet" (Goble 2009). Indeed, Apple's success in the last decade is uncontested, and sales figures confirm that no Saddle had developed following all of its recent products. But this should not be surprising considering that Apple produces its products in a way that avoids the dual-market structure!

Apple excels at optimizing product functionality and convenience and at creating exceptional product experiences. But Apple is *not* a technological innovator. There were MP3 players long before Apple launched its iPod, many smartphones long before the iPhone, and e-book readers and tablets before the iPad. The modifications that Apple makes to the basic innovation, however, are considered, at least to its clients, masterpieces, which include a critical mass of perceived differences. As a result, consumers view Apple products as belonging to a category of their own. This is why new Apple products *seem* innovative. If Apple's flagship successes were genuine innovations, however, its penetration product would have been much slower and paved with many more failures, and Saddles. This does not happen because Apple designs products for the main market. The simplicity and intuitiveness of its devices are mandatory. For example, the iPhone never had a help function. It was so simple, so intuitive, that even main-market consumers could learn how to use it on the spot.

This means that Apple is a social, rather than technological, innovator. Apple products lead social revolutions through outstanding product design and experience. But current Apple products (such as the iPhone and iPad) are not really radical innovations, and, accordingly, they do not elicit different degrees of resistance across different consumer populations. As a result, the early and main market adopt almost simultaneously. The iPhone had such wide appeal because it imposed no adoption costs. People already knew how to use cameras, music players, and cell phones. All the iPhone did was combine existing features into a single sleek design. Products that are both technologically innovative and avoid a Saddle do exist, but this does not seem to be the case for Apple products. Apple avoids the Saddle because its best-selling products evoke little main market resistance.

Undeniably, one of the secrets behind Apple's success is the Apple community. This large group of loyal customers nevertheless does not simply embrace anything Apple has to offer. Recall that Apple also had failures in the past (the Cube TV, for example), like any other innovative firm. But members' almost religious support of Apple effectively erodes resistance and provides a large safety net of main-market consumers who will seriously consider any new product that Apples introduces.

Apple also creates a new type of dialogue connecting manufacturer, developers, and main-market consumers. Apple community members are heavily involved in product improvements and developments. The App developer community has grown in leaps and bounds, and the potential benefits of successful Apps have enticed even main-market consumers to participate in product development. Not only have these user-developed iPhone Apps substantially increased the product's value; their public and acclaimed role in the development process has, perhaps for the first time, also created a bridge between early-adopting App developers and main-market product users. This Apps phenomenon has now spread to all smartphones and tablets.

When Apple does launch products that are truly technologically innovative, the results are more in line with what innovation adoption studies typically show. The Apple Pippin, for example, launched in 1995, competed with other leading game consoles of the period, but at double the price, and was soon shelved. Another Apple product that is considered a failure is Apple TV. Interestingly, Apple TV can be considered a genuine technological innovation, aiming to merge TV and the Internet. But it was launched with many technological limitations, and, even worse, it restricted the type of content that could be played, which made the device viable for only a fraction of all users.

Today, however, the PC/TV convergence market offers new generations of streamers and Blu-ray players a multitude of playback and media purchasing options. It will be interesting to see whether Apple picks up the gauntlet and carries Apple TV across the chasm of main-market resistance by incorporating the lessons learned from the early adopters in this category.

References

Casti, John L. 1989. *Alternate Realities: Mathematical Models of Nature and Man*. New York: John Wiley and Sons.

Chandrasekaran, Deepa, and Gerard Tellis. 2006. Getting a Grip on the Saddle: Cycles, Chasms, or Cascades? PDMA Research Forum, Atlanta, October 21–22.

———. 2011. "Getting a Grip on the Saddle: Chasms or Cycles?" *Journal of Marketing*, forthcoming.

Electronics Industries Association. 1976. *Electronic Market Data Book*. Washington, DC: Electronics Industries Association.

Goble, Gordon. 2009. "Apple's Worst Products and Biggest Failures." Digital Trends, August 25, http://www.digitaltrends.com/features/apples-worst-products-and-biggest-failures/.

Goldenberg, Jacob, Barak Libai, and Eitan Muller. 2002a. *Is the Bandwagon Rolling?: The Chilling Effect of Network Externalities on New Product Growth*. Tel Aviv University Faculty of Management, the Georges Leven High-Tech Management School.

———. 2002b. "Riding the Saddle: How Cross-Market Communications Can Create a Major Slump in Sales." *Journal of Marketing* 66 (2): 1–16.

Goldenberg, Jacob, Barak Libai, Eitan Muller, and Renana Peres. 2006. "Blazing Saddles: Early and Main Markets in Product-Life-Cycle in High-Tech Industries." *Economic Quarterly* 53: 249–71.

Golder, Peter N., and Gerard J. Tellis. 1993. "Pioneer Advantage: Marketing Logic or Marketing Legend?" *Journal of Marketing Research* 30 (May): 158–79.

———. 2004. "Growing, Growing, Gone: Cascades, Diffusion, and Turning Points in the Products Life Cycle." *Marketing Science* 23 (2): 207–18.

Lehmann, Donald, and Mercedes Esteban-Bravo. 2006. "When Giving Some Away Makes Sense to Jump-Start the Diffusion Process." *Marketing Letters* 17 (4): 243–54.

Lewis, Geoff. 1985. "Will Cheap IBM Clones Pep Up a Sluggish Market?" *Business Week*, December 2, 142d–143d.

Mahajan, Vijay, Eitan Muller, and Rajendra K. Srivastava. 1990. "Determination of Adopter Categories Using Innovation Diffusion Models." *Journal of Marketing Research* 27 (1): 37–50.

Manjoo, Farhad. 2010. "What Are You Waiting For? You're Not an Idiot for Buying an iPad on April 3." *Slate*, March 16, http://www.slate.com/id/2248020/.

Moore, Geoffrey A. 1991. *Crossing the Chasm: Marketing and Selling Technology Products to Mainstream Customers*. New York: HarperBusiness.

Muller, Eitan, and Guy Yogev. 2006. "When Does the Majority Become a Majority? Empirical Analysis of the Time at Which Main Market Adopters Purchase the Bulk of Our Sales." *Technological Forecasting and Social Change* 73 (9): 1107–20.

Murphy, Brian J. 1987. "Long Live the Home Market (Whatever It Is)." *High-Tech Marketing*, June, 20–27.

Parker, P. M. 1994. "Aggregate Diffusion Forecasting Models in Marketing: A Critical Review." *International Journal of Forecasting* 10 (2): 353–80.

Rogers, Everett M. 1962. *Diffusion of Innovations*. New York: Free Press of Glencoe.

———. 1995. *Diffusion of Innovations*. 4th ed. New York: Free Press.

Sultan, Fareena, John U. Farley, and Donald R. Lehmann. 1990. "A Meta-Analysis of Applications of Diffusion Models." *Journal of Marketing Research* 27 (1): 70–77.

Tellis, Gerard J., and Peter N. Golder. 2001. *Will and Vision: How Latecomers Grow to Dominate Markets*. New York: McGraw-Hill.

Van den Bulte, Christophe, and Yogesh V. Joshi. 2007. "New Product Diffusion with Influentials and Imitators." *Marketing Science* 26 (3): 404–21.

Epilogue

"Change is great; you go first," as Dilbert says. Innovation and resistance go hand in hand. Resistance is often a very sensible reaction. Our aim in this book was, however, to discuss the various factors that could elicit or exacerbate resistance and demonstrate several ways in which resistance manifests itself in market behavior. When explaining resistance, we discussed factors that reside within the individual along with characteristics of the particular situation in which the marketing of innovation takes place. This latter group includes features of the innovation that may threaten some consumers, characteristics of the manner in which the innovation is introduced and promoted, and attributes of the context within which this is done. The model we presented in the introduction, and developed throughout part I, provides an integrative overview of these various sources of resistance and of how they interact in creating consumers' resistance to innovation.

As our model suggests, the most immediate and basic reason individuals have for resisting innovations is the potential threats the innovation causes. Everything else constitutes the setting, which can increase or decrease one's perceived threat. The context within which innovations are introduced, the manner in which they are presented, and even consumers' personality may all contribute to one's sense of possible threat. For those interested in alleviating or overcoming resistance, the main objective is to assuage such perceived threats, the key to which is acknowledging their existence and their exacerbating factors. Rather than giving up on resistors and blaming them for being irrational, we propose to try to view innovations from their perspective.

Similar to the typical view in marketing, organizational practitioners and researchers tended to view resistance as a sign of employee irrationality. It was taken for granted that management knows what it's doing and initiates the necessary and well-planned steps when introducing a change. For the organization to succeed, all that was presumed to be necessary was that employees accept and fully cooperate with the change. A number of scholars have challenged this view, however, and highlighted the notion that resistance reflects employees' legitimate concerns, which often go beyond their personal needs and involve their care for the organization. Viewed as such, resistance is considered an important warning that should be heeded and from which lessons should be learned when designing change. Corresponding with our arguments in chapter 3, this notion stresses the importance of involving innovation recipients in the planning and launch procedures of the innovation. In this sense, potential resistors assist firms in circumventing resistance. Insights from part II of the book suggest additional ways in which past resistance can be used to overcome future resistance.

In chapter 5, we discussed the leapfrogging effect and the consequence that Laggards' leaps, through their word of mouth, have on products' adoption curves. Our approach to estimating this effect was conservative, whereby the effect of a Leaping Laggard was assumed to be the same as that of an "ordinary" Innovator. Based on what we've learned from the Saddle effect, however, one could argue that Leaping Laggards would have a more substantial effect. The effect of learning that someone who typically resists innovations had purchased the latest technology, despite the costs involved in its adoption (financial and otherwise), is likely to be even stronger than observing a main-market consumer do the same.

This corresponds with our discussion in chapter 7, about how information that comes from adopters who were inclined to resist the innovation may be much more valuable and influential than that of early adopters who adopt blindly. For example, typical Innovators may be more willing to compromise on design imperfections for the thrill of getting something new. They may also be more forgiving toward such imperfections because they may feel sufficiently skilled in dealing with them. As we argued in chapter 7, however, main-market consumers, who are more reserved in their response to innovations, may actually provide richer product feedback that requires firms to work harder on their products' development. R&D expenses may be higher, but products end up better, more sustainable, and in the long run often yield greater sales. Resistors may therefore be a blessing in disguise, and can be considered an asset for firms in ultimately speeding up the adoption of their products.

Of course, sometimes it is harder to see the silver lining in resistance. In chapter 6, we discussed the phenomenon of negative word of mouth. An interesting case is when increasing marketing expenditures (for example, advertising) can lead to increased losses. This can occur when the early activation of resistance leads to market shrinkage. In this case, the solution is to adjust marketing efforts to the anticipated resistance. Another way to look at it is that resistance forces companies to employ marketing efforts that are more carefully targeted toward a specific audience, thus increasing their efficiency and indirectly improving marketing campaigns.

The typical negative outlook on resistance as a nuisance is clearly too simplistic. As noted above, for example, because of their tendency to lag, highly resistant consumers may become a valuable marketing asset. Even setting aside the possibility that Leaping Laggards will have a stronger effect than that of typical Innovators, we demonstrated in chapter 5 the potential impact that resistors could have on improving firms' NPV. But even more broadly, we suggest that innovation and resistance cannot be separated, even logically; the two concepts help define each other.

Imagine a marketer's ideal world—a world with no resistance—in which consumers are keen to adopt anything new. What would an innovation mean then? Without the obstacles and barriers, firms would swamp markets, probably with less sophistication, less originality, and lower standards. This marketer's dream would likely become the consumers' nightmare. Either consumers would suffer or they would gradually develop some evaluation criteria that would ultimately elicit resistance.

Resistance serves as a gatekeeper. Society includes consumers and firms, and in any developed society, firms successfully develop and market innovations. Resistance is an essential component to maintaining equilibrium and to channeling firms to develop sustainable and meaningful innovations. Resistance thus enhances the quality of innovations. At the same time, it does not prevent progress. Throughout our formulations in this book, we have discussed many cases of resistance, but the resistance described was typically local or temporary. Despite the initial objections to the invention of the phone, we all use phones.

Unfortunately, our analyses suffer from the problem of sample selection. Obviously, we cannot examine cases in which too much resistance may have led industries to unduly "pull the plug" on innovations that could have meaningfully improved our lives. We can only assume that such cases exist and acknowledge that this may be the price of benefiting from the quality control that resistance provides.

FIGURE E.1. What is this "thing"? Photographer unknown.

This does not mean we should idly accept resistance and refrain from addressing it. It simply means that we should more readily accept its productive functions. Consider, for example, figure E.1. Could you guess what this is?

This is the ancestor of our flash drives. It is the volume and size of 5 MB of memory storage in 1956. In September 1956, IBM launched the 305 RAMAC, the first computer with a hard disk drive (HDD). The HDD in figure E.1 weighed over a ton.

Let us assume that size and weight were not disadvantageous and would not thus provide potential adopters a reason to resist the innovation. Would firms have bothered to improve it? The personal computer would have been eventually developed, but if not for resistance, due to a variety of reasons, it could very well have dawned much later. Resistance is a key trigger for the development of improvements. It is an inevitable and essential component on the way to progress.

* * *

In this book, we focused on the roots of resistance and exemplified some of its manifestations in market behavior. This market behavior, as we have shown, follows from the individual-level sources of resistance, aggregated to the market level, developed over time. Our emphasis was on the conceptual framework that explains resistance and follows its consequences, and while we proposed several means of relieving it, we leave it to experts in the field to translate our scientific insights into practical actions in marketing.

Index